Mohammed Faraj Saeid
Sim Yee Chin

Production of n-butanol from the Hydrogenation of Butanal

Case Study and Simulation on the Effect of Different Reactor Scheme and Operating Conditions

Noor Publishing

Imprint

Cover image: www.ingimage.com

Publisher:
Noor Publishing
is a trademark of
Dodo Books Indian Ocean Ltd., member of the OmniScriptum S.R.L Publishing group
str. A.Russo 15, of. 61, Chisinau-2068, Republic of Moldova Europe
Printed at: see last page
ISBN: 978-620-4-72063-0

ACKNOWLEDGEMENTS

My deepest gratitude goes to my beloved parents also to my sisters and brother for their deeply love and endless support, prayers and encouragement. To those who indirectly contributed in this research, your kindness means a lot to me. Thank you very much.

ABSTRACT

Butanol is one of the most promising biofuels with many advantages over bioethanol, and it has been used in large-scale processes over the last decades. Butanol plant using technology of butanal hydrogenation by using three reactors in the series. This study was aimed to simulate butanol production from hydrogenation of butanal by using Aspen Plus. The simulation was performed to evaluate the impact of by-passing the second liquid phase reactor in the series due to the catalyst blockage problem leading to the plant shutdown. RPlug reactor model incorporated with NRTL thermodynamic model and kinetic equations was validated. The simulation results deviated <10% from the plant data. The attested model was subsequently used to simulate butanal hydrogenation at different reactor configuration, reactor operating conditions and reactor size. The targeted final total conversion was 99.5% of butanal. The butanal conversion increased with the increase of reaction temperature and residence time. On the other hand, the reactor pressure only affected the performance of the vapour phase reactor but not the liquid phase reactor. In comparison to the original reactor configuration (a vapour phase reactor followed by two liquid phase reactors), best reaction parameters enabling the attainment of total conversion of 99.5% are 213°C, 158.5°C, and 159.5°C of the reaction temperature and 21, 16, and 16 bars of the reaction pressure. While the best reaction parameters in modified configuration I (a vapour phase reactor followed by single liquid phase reactor) are 255°C and 166.5 °C of the reaction temperature and 21 bar and 16 bar of the reaction pressure for achieving the targeted total conversion of 99.5%. While the best reaction parameters in modified configuration II (single vapor phase reactor) to achieve the final total conversion of 99.99% are 285°C of the reaction temperature and 30 bar of the reaction pressure which is required to exceed the limits of the industrial plant operating conditions. Best reactor diameter and reactor length values in the three reactor schemes were determined as 2.75 m and 10.92 m. Conclusively, the idea of bypassing liquid phase reactors for hydrogenation of butanal in the series could be adopted to achieve the targeted total conversion, provided the plant is allowed to operate a more severe operating conditions, considering the safety factor.

TABLE OF CONTENT

LIST OF SYMBOLS

Z	Axial distance from reactor inlet, ft
E	Activation Energy
u	Aldehyde
C_c	Coolant heat capacity, Btu/Lb/°C
D_p	Catalyst pellet diameter, ft.
e	Catalyst bed void fraction
T_c	Coolant temperature, °C
A	Constant of the Arrhenius equation
C_p	Gas heat capacity, Btu/LbMol/°C
F	Gas flow-rate, LbMols/Hr
R	Gas constant, 1.987 Cal/Gmol°C
ρg	Gas density, Lb / ft^3
ΔH_{rj}	Heat of reaction for reaction j, Btu/LbMol
MODE	Heat transfer mode: -1 for coolant countercurrent to gas 0 for isothermal shell-side +1 for coolant co-current with gas
K-£	Kinetic parameter
Y-£	Mole fraction of component i
η	Number of tubes in reactor
r_j	Net rate of reaction for component i, LbMols/Ft^3/Hr
U	Overall heat-transfer coefficient, Btu/Ft^2/Hr/°C
K_o	Pre-exponential rate constant
P_i	Partial pressure of component i, atmospheres
E	Reaction activation energy, Cal/Gmol
K_{eq}	Reaction equilibrium constant (dimensionless)
R_i	Reaction rate for reaction j, LbMols/Ft^3/Hr
T_r,T	Reaction temperature, °C
βj	Relative activity factor for reaction j
Vo	Superficial gas velocity, ft/sec.
α_{ij}	Stoichiometric coefficient for component i in reaction j
D	Tube diameter, ft.
V_0	Volumetric flow rate, L/min

LIST OF ABBREVIATIONS

H	Hydrogen
IB	i-butanal
IOH	i-butanol
MW	Mean molecular weight of gas, Lb/LbMol
MW-£	Molecular weight of component i, Lb/LbMol
NB	Mole flow rate, Kmol/hr
N	*n*-butanal
NOH	*n*-butanol
PBR	Packed Bed Reactor
P	Pressure, bar
T	Temperature, °C

CHAPTER 1

INTRODUCTION

1.1 Background of study

The alcohol, *n*-butanol is an organic substance being proposed as an alternative fuel for internal combustion engines. Currently, butanol is produced chemically by either the oxo process starting from propylene (with H_2 and CO over rhodium catalyst) or the aldol process starting from acetaldehyde (Qureshi et al., 2014). And its combustion characteristics, close to those of gasoline and diesel, grow attention to its viability as a renewable and economically interesting substitute to petrol-based fuels(Alviso 2015).

The use of fuels, specially in transportation and industrial processes, is seen as one of the most promising energy solutions for the future. Attention to this field is driven by uncertainties related to oil prices, greenhouse gases and pollutant emissions, as well as the need for lower dependence from petrofuel producers and higher diversity in sources (Jin et al., 2011).

According to the Key World Energy Statistics 2014 (IEA, 2014), 81,7% of the world primary energy supply comes from fossil fuels, being 31,4% of the total amount represented by oil. fuels and waste, including firewood - one of the main energy sources in developing nations - account for only 10%. This is a critical situation, not only because of the dependence from fossil fuel productors but also because of the huge amount of pollutants and greenhouse gases emitted in the atmosphere. Use of butanol as fuel will contribute to clean air by reducing smog-creating compounds, harmful emissions (carbon monoxide) and unburned hydrocarbons in the tail pipe exhaust (Ezeji et al., 2006) .

Alcohols have also high octane numbers, which make them suitable for spark ignition engines. Some countries already consume alcohols as a fuel in large scale, being Brazil and the United States the most important examples, using it pure or blended with gasoline in different proportions (Agarwal 2007).

Roughly 2 x 10^6 tons of butanol are produced annually, for use as a plasticizer, an industrial solvent, and an intermediate in the production of butyl acetate. The demand for butanol is expected to increase in the future as a consequence of recent studies showing that butanol is a viable alternative to ethanol as an additive to gasoline, a key ingredient in lacquers and varnishes. While many processes exist for the production of butanol, such as the aldol condensation of ethanal, oxidation of butane, or the enzymatic fermentation of sugars, the overwhelming majority of butanol is produced in a three-step process involving the homogeneously catalysed hydroformylation of propene. Separation of the resulting butanal, and subsequent hydrogenation of butanal. Separation of butanal is necessary because aldehyde hydrogenation catalysts such as supported Pt and Pt–Sn, Rh–Sn, and Ni are poisoned by CO, as well as being active for alkene hydrogenation (Hanna et al., 2014).

The Oxo synthesis (hydroformylation) process is based on the reaction of propylene with Oxogas to give a crude mixture of *n*- and iso-butyraldehyde, aldehydes are reacted with hydrogen to form butanol via hydrogenation and distillation (Walczuch 2013). The hydroformylation reactions occur in the presence of cobalt or rhodium complexes with different ligands, recognized as homogeneous catalysts. Nowadays, rhodium/TPP is the most used hydroformylation catalyst. The existent industrial plants are designed on two major technologies, available commercially: gas recycle technology and liquid recycle technology. In both technologies, the gas-liquid reactions take place in continuous stirred tank reactors (Tuţă and Bozga 2011). The adoption of a low-pressure rhodium-based catalyst system in place of high pressure cobalt for the hydroformylation of propylene by reaction with carbon monoxide and hydrogen to produce butanal (an 'oxo' reaction) has brought large cost benefits to oxo producers. The benefits derive from improved feedstock efficiency, lower energy usage and simpler and cheaper plant configurations (Tudor and Ashley 2007). The butanal was then hydrogenated to butanol through the high pressure and low pressure hydrogenation packed bed reactors.

1.2 Case study

A butanol production plant adopts the technology of using three packed bed reactors in series (1 vapor phase reactor and 2 liquid phase reactors) for the high pressure and low-pressure butanal hydrogenation respectively. This butanol production plant frequently encountered the catalyst blockage problem in the third reactor, the Low-Pressure Hydrogenation (LPH) reactor that caused a plant shutdown. In view of minimising the production and revenue loss due to the plant shutdown, the plant operators propose to continue producing the demanded product with the two precede hydrogenation reactors in series, by-passing the third reactor. The elimination of the low-pressure hydrogenation step would have some effects on the product quality, in terms of conversion, selectivity and yield. As no experimentation is allowed in industrial plants, an accurate process simulation results are of paramount to validate the proposal. Simulation is a process of designing an accurate operational model of a reactor and conducting investigations with this model for the purpose either of understanding the behaviour of the system or of evaluating alternative strategies for the development or operation of the system(Alexandre et al. 2014). The present study will be developing an accurate process model for the simulation of butanal hydrogenation for the production of butanol. The process with different operating conditions and reactor schemes will be simulated. The simulation results will then be used to support the initiative to minimise the plant down-time.

1.3 Aim of the study

The aim of the present work is to simulate and investigate the effect of the reactor scheme and operating conditions to the reaction performance. The industrial plant data was used to validate the model. Three reactor configurations of butanal hydrogenation were considered for the simulation analysis. Original reactor configuration (single vapour phase reactor followed by two liquid phase reactors), modified configuration I (single vapour phase reactor followed by single liquid phase reactor), and modified configuration II (single vapour phase reactor).

CHAPTER 2

TECHNIQUES IN THE PRODUCTION OF N-BUTANOL

2.1 Introduction

The present chapter includes the market and demand of butanol that justify the requirement of an optimum continuous production process with minimum unplanned shut down. Different types of production technologies are then reviewed before the focus is shifted to the Oxo synthesis process. The type of reactor and catalysts used in the butanal hydrogenation for butanol production are evaluated. The corresponding operating conditions are also reported. Subsequently, the importance of the simulation study for the butanal hydrogenation is marked by the findings in some of the simulation studies for the relevant processes.

Butanol is a chemical that has excellent fuel characteristics. It contains approximately 22% oxygen, which when used as a fuel extender will result in more complete fuel combustion. Butanol is an important intermediate and solvent in the chemical industry. Moreover, compared to ethanol, butanol has higher energy density and lower vapor pressure, so butanol is also considered a preferred fuel additive or even a potential replacement for gasoline; the use of butanol as fuel will contribute to clean air by reducing smog-creating compounds, harmful emissions (carbon monoxide) and unburned hydrocarbons in the tail pipe exhaust (Acres 2007).

Butanol is used mainly as an industrial intermediate to make chemicals: primarily butyl acrylate, but also butyl acetate, ethylene glycol, and butyl xanthate among others (Acevedo et al., 2010).

2.2 Market and Demand of Butanol

n-Butyl alcohol or *n*-Butanol or Normal Butanol is a primary alcohol with a 4-carbon structure and the molecular formula C_4H_9OH. Its isomers include iso butanol, 2-butanol, and tert-butanol (Liu 2012).

According to SRI statistics, the world largest *n*-butyl alcohol manufacturer now is BASF, the largest diversified chemical company in the world and being headquartered in Germany. With the rapid downstream development of OXO in China over the years, the supply of OXO falls short of demand. In the recent years, the high profit margin of OXO products stimulate the investment enthusiasm to OXO device in domestic market. As one of the main materials, propylene has extensive sources. As another main material, synthesis gas has mature production technology. The supply of propylene and synthesis gas provides material guarantee for the construction of OXO device. The obtaining of process technology is not exclusive, which offers feasibility technically (Liu 2012).

2.2.1 World Market

2.2.1.1 Main Manufacturers

According to SRI statistics, the world largest n-butyl alcohol manufacturer now is BASF with the production capacity of 649kmt/y. The total production capacity of top 10 manufacturers is 2,775kmt/y. See details in Table 2.1 (Liu 2012).

Table 2.1 Global top 10 manufacturers of n-Butyl Alcohol kmt/y

No.	Company	Capacity
1	BASF	649
2	Dow Chemical Company	526
3	Oxea Group	280
4	Formosa Plastics Group	250
5	Eastman Chemical Company	247
6	CNPC	195
7	Petronas	190
8	Sasol Limited	188
9	Sasol Limited	130
10	SINOPEC	120
	All Others	793
	TOTAL	3568

Source: Liu (2012).

2.2.2 Actuality of Supply and Demand

According to SRI statistical data, the production capacity of world *n*-butyl alcohol is 3,568,000 tons/year, the output is 2,944,000 tons/year, the operating rate is 82.5% and the apparent consumption is 2,944,000 tons/year in 2010. The top 3 regions with large production capacity are Asia, North America and Western Europe, which respectively take up 38%, 31% and 18% in world total production capacity. The most important consumer is Asia, whose consumption takes up about 53% of world total consumption. And then there are North America and Western Europe, whose consumption respectively takes up 22% and 19%. Both production and consumption gradually focus on Asia. See the supply and demand of world OXO market in 2010 in Figure 2.1 (Liu 2012).

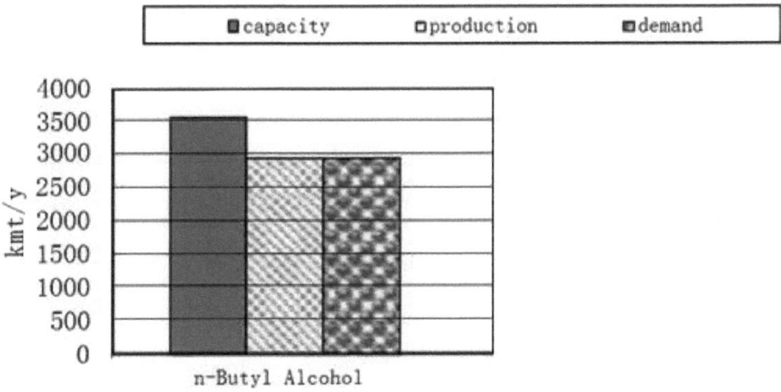

Figure 2.1 Supply and Demand of World OXO Market in 2010

Source: Liu (2012)

According to SRI forecast, the production capacity of world *n*-butyl alcohol will be 4,517,000 tons/year, the output will be 3,824,000 tons/year, the operating rate will be 84.7% and the consumption will be 3,824,000 tons/year in 2015. The production capacity of world *n*-butyl alcohol in 2020 is preliminarily predicted as 4,727,000 tons/year and the quantity demanded is 4,233kmt (Liu 2012).

2.3 Butanol Production Processes

Currently, butanol is produced chemically by either the oxo process starting from propylene (with H_2 and CO over rhodium catalyst).Butanol is also produced by fermentation of corn and corn-milling byproducts (fermentable sugars) (Acres 2007).

2.3.1 Fermentation Process

Acetone-Butanol-Ethanol (ABE) fermentation is an important bioprocess that produces *n*-butanol. It was first discovered by Pasteur in 1861 (Jones et al.1986). Before the 1950s acetone, butanol and ethanol (ABE) fermentation by Clostridium acetobutylicum was the main route to produce butanol. This process began to decline due to the increasing of substrate cost and the appearance of a new, more economical, petrochemical route. Today, almost all butanol is produced from petrochemical feedstock, although in some countries production continued through the fermentation route (Dürre, 1998). The continuous increasing of greenhouse gas emissions and the resulting concerns about climate change, together with the availability of new renewable feedstocks, are leading to a growing interest in sustainable industry and, consequently, to a renewed interest in ABE fermentation (Lodi et al. 2016).

ABE fermentation is performed at temperatures between 27-37 °C and 1 atm by a large variety of Clostridia strains that are able to ferment different sugars, including glucose, xylose, arabinose and mannose (Ezeji and Blaschek, 2008).

However, the microorganisms used in ABE fermentation suffer from product inhibition, mainly from butanol. Typically, 20 g/L of total solvents, with a butanol concentration as low as 13 g/L, are achieved in the bioreactor during a batch process (Maddox, 1989), restricting the sugars concentration in the substrate to about 60 g/L. These limitations deeply affect the costs of the process, both CAPEX due to the need of large process volumes and OPEX involved in downstream separations for product recovery, usually performed by distillation (Lodi et al.2016).

The application of an in-situ recovery technique, able to remove butanol from the fermentation broth during fermentation, currently appears to be the most viable solution to solve the problems related to ABE fermentation and to improve the profitability of the process (Ezeji et al.2005).

Product removal techniques include gas stripping (Groot et al.1989), liquid-liquid extraction (Barton and Daugulis, 1992), adsorption (Nielsen and Prather, 2009) and pervaporation (Matsumura et al.1988).

Gas stripping has several advantages in comparison with the other technologies because it is simple to operate and to scale up, it does not require expensive equipment or chemicals, it does not remove nutrients and reaction intermediates from the broth and it is not harmful to cells. (Ezeji et al. 2003). In gas stripping, an inert gas is contacted with the fermentation broth, capturing the solvents, and is then passed through a condenser. Here the stripped components condense, while the gas can be recycled to the stripping section (Lodi et al. 2016).

The application of gas stripping results in the use of concentrated sugar solutions in the fermenter (Qureshi and Blaschek, 2001), in a reduction in butanol inhibition and in higher sugar utilization (Maddox and Qureshi, 1995), reducing process volumes. The concentration of the butanol in the recovered stream is higher than in the fermentation broth.

However, gas stripping usually also removes a large amount of water with butanol and requires a higher energy input, because of its lower butanol selectivity if compared to other separation techniques (Oudshoorn et al.2009; Qureshi et al. 2005; Vane, 2008).

In fact, by removing toxic solvents from the culture at a rate similar to that at which they are formed, it should be possible to ferment more concentrated sugar solutions, reducing capital costs, because of the higher fermenter productivity, and operating costs for product recovery (Maddox, 1989).

2.3.1.1 Process Description

In the gas stripping process, fermentation products are removed from the broth by contacting it with an inert gas stream. Either nitrogen or the fermentation gases themselves (CO_2 and H_2) can be used for this operation (Groot et al.1989; Ezeji et al. 2003). Attention should be paid in the use of fermentation gases, since carbon dioxide could solubilise in the fermentation broth and decrease its pH, a parameter that plays a key role during ABE fermentation (Maddox, 1989; Dahlbom, 2011). Therefore, nitrogen has been preferred as stripping agent. Gas stripping can be performed either bubbling the gas through the fermenter or in an external column (Groot et al.1989; Ezeji et al. 2003). An external stripper represents a more appropriate design, because of the higher performances of a multistage unit compared with a single equilibrium stage. Moreover, with this configuration it is possible to use nitrogen as a stripping agent while maintaining a beneficial fermentation gas, positive head space pressure (Lodi et al.2016).

In the studied configuration the fermentation broth is withdrawn from the fermenter and sent to a stripping column: here the ABE mixture is stripped with nitrogen, passing from the liquid phase to the vapour phase. The fermentation broth is then recycled to the reactor, while the vapour phase is sent to the condenser. Here the stripped compounds are condensed and the regenerated nitrogen is recycled to the stripping column. A nitrogen make-up is required to compensate the amount of gas that remains dissolved in the product stream. The condensed ABE mixture has a higher butanol concentration than the initial fermentation broth. If the global composition of condensate falls in the liquid-liquid demixing region, a phase separation occurs, giving an organic phase rich in acetone, butanol and ethanol and an aqueous phase rich in water. The condensate is collected in a decanter in which the two liquid phases can be separated and sent to a distillation train for further separation. The lean fermentation broth exiting from the stripping column can be recycled to the fermenter. A schematic representation of the proposed process is shown in Figure 2.2 (Lodi et al. 2016).

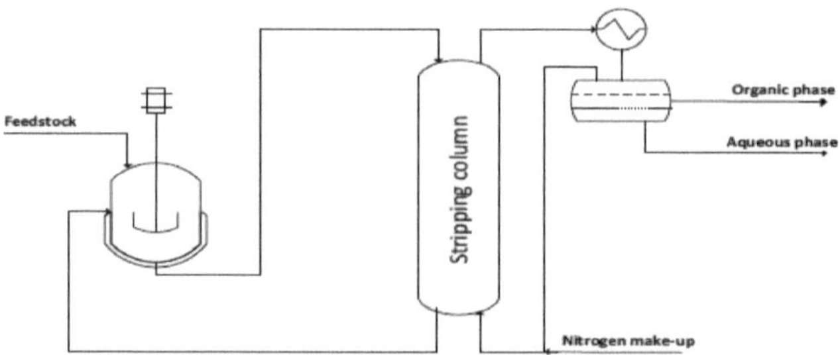

Figure 2.2 Process flow diagram of the integrated stripping unit for recovery of butanol from fermentation broth

Source: Lodi et al. (2016).

2.3.2 Oxo -C4 Process

The Oxo-C_4 process is based on the reaction of propylene with oxogas or syngas to give a crude mixture of *n*- and iso-butanal. The aldehydes are separated and purified by distillation. *n*-butanal is processed in a series of enalization, hydrogenation and distillation systems to give 2-ethylhexenol. Simultaneously, aldehydes are reacted with hydrogen to form butanols via hydrogenation and distillation. The Oxo-C4 process including a hydrogenation catalyst, which is a BASF patented technology.

Syngas (synthesis gas) is a mixture of carbon monoxide (CO) and hydrogen (H_2), which is the product of high temperature steam or oxygen gasification of organic material such as biomass. While propylene is a colorless gas and produced in vast quantities by the petrochemical industry by the cracking of propane or higher hydrocarbons in the presence of steam under conditions very like those outlined for ethylene (Martin 2005).

Propylene is an important industrial chemical used as a feedstock in the production of polypropylene, acrylonitrile, propylene oxide, 2-propanol, alkylated fuels, and other chemicals. It is obtained from steam cracking in pyrolysis furnaces in olefins plants and in refinery catalytic-cracking reactors. It is a colorless gas and has anesthetic properties at high concentrations. Propylene is usually transported in the liquid state at high pressures (Sanders 1977).

Briefly, the process consists of the following production steps:

- Oxo-C4 Synthesis
- Aldehyde Distillation
- 2-ethylhexenol Production
- *n*-butanol production; and
- iso-butanol Production

2.3.2.1 Oxo-C4 Synthesis (Hydroformylation)

The hydroformylation reaction was first reported by Dr Otto Roelen of Ruhrchemie AG, Germany, in 1938 and was given by German researchers the description of 'oxo' synthesis. Alcohols synthesised by the hydrogenation of aldehydes produced from hydroformylation tend therefore to be called oxo alcohols(Matthey 2017).

Table 2.2 Comparison Between Different Operating Conditions on Hydroformylation Processes

	Davy Process Technology		**BASF Process Technology**	
Parameter	**Normal Operating Range**	**Reaction Phase**	**Normal Operating Range**	**Reaction Phase**
Temperature (°C)	85-120	Vapor phase	100	Vapor phase
Pressure (bar)	<11		20	
Rh concentration % wt	200-400		100-300	
TPP concentration % wt	10-12		5-10	

Source: Tuţă and Bozga (2011); Van (2007).

(Van 2007) investigated improvements in or relating to the commercial scale hydroformylation of propylene to produce butyraldehyde and butanol. Liquid propylene is reacted with oxogas in the presence of a soluble rhodium catalyst. The product is a mixture of *n*- and iso-butanal with by-products of propane and high-boiling components. This exothermic chemical reaction is represented by the following equation:

$$C_3H_6 + CO/H_2 \rightarrow C_4H_8O + C_2H_4(CH_3)CHO + by\ product \tag{2.1}$$

Propylene, supplied by the neighbouring propane dehydrogenation plant, is purified in an absorber bed unit before being introduced into the oxo reactor (R100). It is reacted with oxogas that is delivered from the Syngas Plant via pipeline. The reaction conditions are approximately 100°C and 20 bar with the reactor liquid containing 40-70% high boilers, 100-300ppm rhodium and 5-10% triphenyl-phosphine (TPP) (Van 2007).

For recycling of the un-reacted components, the gas stream from the reactor top is fed to the recycle gas compressor (C105) before being sent back to the reactor. A purge stream containing high concentrations of methane, entrained propylene and butanal is taken off from the compressor. It is then passed through a wash tower (T110) that serves as a vent gas scrubber by using cold degassed aldehyde as the washing fluid. Vent gas from the tower is used as fuel gas in the Syngas Plant and the bottom products are sent to the degassing tower (T130) (Van 2007).

The reactor liquid effluent carrying the product is flashed into a drum (D120) and flash tower (T121), where the butanal mixture and C3 (propane and propylene) components are separated from the high-boilers. The liberated gas is collected, compressed and fed back to the oxo reactor inlet (Van 2007).

The aldehyde mixture which still contains propylene and propane is passed from the flash tower (T121) to two degassing towers (T130 and T133) operating in series at 8 and 24 bar respectively, where the C3 components are distilled overhead and sent back to the propane dehydrogenation plant where propylene is recovered and re-used as feedstock (Van 2007).

The propylene-free crude aldehyde mixture from the degassing tower (T130) bottom is sent either to the aldehyde distillation section or to the oxo-C4 tank (TK610) (Van 2007).

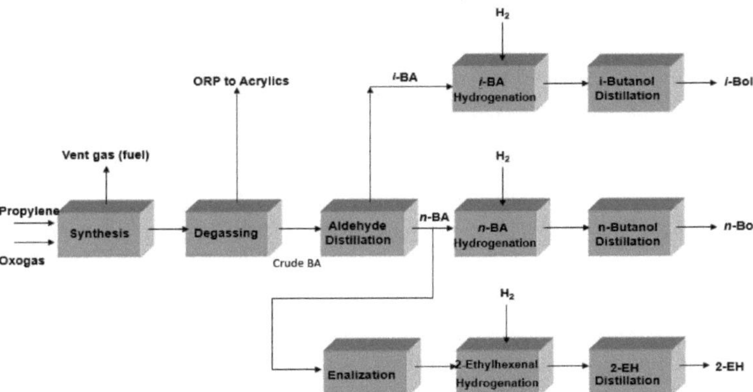

Figure 2.3 A Simplified Process Flow Diagram

Table 2.3 The Influence of Main Operating Parameters on Hydroformylation Process Performance

Parameter	Normal operating range	Parameter variation	Reaction rates	Product *n/i* ratio	Catalyst stability
Temperature °C	85-120	increase	increase		decrease
Pressure bar	<11	increase	increase		
Rh concentration ppm wt	100-300	increase	increase	slightly increase	decrease
TPP concentration % wt	5-10	increase	decrease	increase	increase

Source: Tuţă and Bozga (2011).

2.3.2.2 Aldehyde distillation

The crude aldehyde mixture from the degassing tower (T130) is refined in the aldehyde distillation tower (T220). The overhead stream from the aldehyde stripping tower (T121) is condensed and collected in the overhead drum (D221), while non-condensables are sent to the flare. The overhead product consists of pure iso-butanal, which is pumped to a storage tank (TK625) (Van 2007).

n-butanal is withdrawn at a side vent from the bottom section of the aldehyde distillation tower (T220). It is recovered as the overhead stream of a subsequent knock out drum (D225) and delivered to the storage tank (TK620). The bottom product (ie. Oxo-oil 840) of T200 is routed to the aldehyde heavies tower (T230) where the heavy ends are pumped to the storage tank (TK-2695) at the ISBL Tank Farm (Van 2007).

2.3.2.3 2-ethylhexenol Production

n-butanal from the BA distillation section is utilised in two different process trains for producing *n*-butanol and 2-ethylhexenol separately. Production of the latter substance comprises the enalization reactors, where *n*-butanal is dimerized, a hydrogenation system and three column refining section for production of purified 2-ethylhexenol. The reactions are represented by the two equations below (Van 2007):

$$2C_4H_8O + NaOH \rightarrow C_4H_{14}O + H_2O \qquad (2.2)$$

$$C_4H_{14}O + H_2 \rightarrow C_4H_{17}OH \qquad (2.3)$$

Enalization of *n*-butanal

n-butanal from the storage tank (TK620) and sodium hydroxide are reacted to give 2-Ethylhexenal in the reactor (R300) equipped with a liquid recycle system. The conversion of *n*-butanal is driven to more than 90%. To control the exothermic reaction, the reactor liquid is cooled in an air- cooled heat exchanger (E301). The temperature is controlled at about 100°C. The overhead stream is cooled and collected in the drum (D310) where phase separation occurs. The organic layer at the bottom of the drum flows to the 2-Ethylhexenal storage tank (TK630), and the water phase is diverted to the waste water treatment plant (Van 2007).

2-Ethylhexenal Hydrogenation

Hydrogen from the Syngas Plant and 2-Ethylhexenal (TK630) are piped to the hydrogenation reactors. The reactors (R320, R321 and R322) are charged with BASF catalyst H1-85 and operated at 36 bar, between 110 to 170°C. They are equipped with a liquid recycle (P320A/B) and an external cooler which is the aldehyde distillation tower reboiler (E220-4). 2-EH from the ISBL tank farm is used for degassing the cooled stream in the degassing drum (D332). Crude 2-ethylhexenol is then pumped to the distillation section or via product cooler (E333) to the storage tank (TK635) (Van 2007).

2-Ethylhexenol Distillation

The 2-ethylhexenol refining system comprises the following:

- Foreruns tower
- Finishing tower; and
- Separation tower.

The bottom product is drawn from the degassing drum (D332) to the 2-EH foreruns tower (T340) which is operated at slightly elevated pressure 1.2 bar. This tower separates the light components from the crude alcohol as overheads before being condensed in the foreruns condenser (E341) and collected in a drum (D341). The major portions of the stream are returned to the tower as reflux and the balance is sent to the butanol foreruns tower (T440) for recovery of small amounts of butanol formed in the hydrogenation reactors. The off-gas is sent to the flare (Van 2007).

The light-free bottom take-off passes to the 2-EH finishing tower (T350), where purified 2-ethylhexenol is withdrawn as liquid from sides of the column. Purified 2-ethylhexenol is transferred to the Plasticizer Plant as feedstock or port tank farm via pipeline (Van 2007).

The condensation energy of the T350 overheads in the finishing condenser (E351) is utilised for the generation of 4 bar steam. The condensed alcohol is collected in the overhead drum (D351), where non-condensables are separated and also sent to the flare. The product is sent back to the tower as reflux with a small overhead take-off being routed back to the 2-EH foreruns tower (T340) (Van 2007).

The bottom stream of T350 consists of substantial amounts of 2-ethylhexenol and high boiling components. They are reclaimed as a top product of the heavies tower (T360) which is operated at reduced pressure. The vacuum compressor (C360) discharges off-gas to the flare and the bottom stream (Oxo-oil 800) is sent to the storage tank (TK-690) (Van 2007).

2.3.2.4 *n*-Butanol Production

Butanal is a high-production volume chemical produced exclusively from hydroformylation of propylene. It is a versatile chemical used in the synthesis of diverse C_4–C_8 alcohols, carboxylic acids, esters, and amines. Its high demand and broad applications make it an ideal chemical to be produced from biomass (Ku et al. 2017).

$$C_4H_8O + H_2 \rightarrow C_4H_9OH \tag{2.4}$$

The above equation represents the *n*-butanol production where *n*-butanal is hydrogenated and distilled. The hydrogenation system includes two reactors in series that are loaded with a heterogeneous BASF catalyst. The distillation section consists of a foreruns tower, a finishing tower and a separation tower (Van 2007).

n-Butanal Hydrogenation

The main hydrogenation reactor (R420) is charged with BASF catalyst H1-84 and is operated at 36 bar, between 110 to 170°C. It is equipped with a liquid recycle (P420) and external cooling system. Hydrogen along with *n*-BA is added to the recycle stream. The bottom of the reactor is designed as a phase separator. The unreacted hydrogen withdrawn from the reactor is combined with the bottom liquid and fed to the finishing reactor (R430). The liquid product stream is cooled and collected in the hydrogenation flash drum (D430) before being degassed in the degassing drum (D432). The pressure vent gas is taken off and used as fuel in the Syngas Plant. The off-gas from the subsequent low pressure hydrogenation degassing drum is burned in the flare. Crude *n*-butanol from D432 is pumped either to the distillation section (T440) or to a storage tank (TK640) (Van 2007).Table 2.4 summarize a comparison of different technologies of hydrogenation Process

Table 2.4 Comparison Between Different Technologies of Hydrogenation Process

	Davy Process Technology			**BASF Process Technology**	
Parameter	**Normal Operating Range**	**Reaction Phase**	**Parameter**	**Normal Operating Range**	**Reaction Phase**
Temperature (°C)	175	Vapor Phase	Temperature (°C)	110-170	Liquid Phase
Total pressure in kg/cm^2 absolute	9.0		Total pressure (bar)	36	
Conversion (%)	82		Conversion (%)	65	
Selectivity to *n*-butanol (%)	99		Selectivity to *n*-butanol (%)	-	
Catalyst a mixture of CuO and ZnO (%) wt	65 , 35		BASF catalyst H1-84 (Kg/day)	164.38	

Source: Van (2007); Bradley et al.(1991).

n-Butanol Distillation

The *n*-butanol refining system also comprises the usual scheme as follows:

- Foreruns tower

- Finishing tower, and
- Separation tower.

The foreruns tower (T440) separates light components (mainly water traces of *n*-butanal) from crude alcohol. They are condensed (E441) and collected in the overhead drum (D441). The organic phase is pumped back to the tower as reflux. The aqueous phase is returned to the hydrogenation reactor (R420) via the feed pump (P621) (Van 2007).

The lights-free bottom product stream passes to the finishing tower (T450), where purified *n*-butanol is obtained as a liquid. The product is cooled and pumped to the storage tank (TK645) in the ISBL tankfarm. Liquid alcohol is collected in the overhead drum (D451), where non-condensables are separated and sent to the flare. The liquid product is pumped back to the tower as reflux and to the foreruns tower (T440) as a feed (Van 2007).

The bottom stream from T450 still contains substantial amounts of *n*-butanol and the heavy ends. *n*-butanol is reclaimed as a top product of the subsequent separation tower (T460), which is operated at reduced pressure and collected inside the overhead drum (D461). The vacuum compressor (C360) discharges its off-gas to the flare. The products are sent to the storage tank (TK-640) in the ISBL tankfarm (Van 2007).

2.3.2.5 Iso-Butanol Production

A single reactor is utilised in the production of iso-butanol. The distillation involves a pre-run tower and finishing tower. *n*-butanal is hydrogenated and distilled to give high purity iso-butanol as shown by the equation below (Van 2007):

$$C_2H_4(CH_3)CHO + H_2 \rightarrow C_2H_4(CH_3)CH_2OH \qquad (2.5)$$

Iso-Butanal Hydrogenation

The *i*-butanal hydrogenation system consists of a single reactor (R520) with a liquid recycle system. The reactor is loaded with BASF hydrogenation catalyst H1-84 and is operated at 36 bar and a temperature range of 110 and 170°C. *i*-butanal along with hydrogen is added to the recycle stream. The two-phase reactor effluent is cooled and flashed in the flash drum (D530) before being degassed in the degassing drum (D532). The pressure vent gas liberated in this drum is used as fuel gas in Syngas Plant and for the subsequent drum (D532), it is burned in the flare (Van 2007).

Iso-Butanol Distillation

The iso-butanol refining system consists of a pre-runs tower and finishing tower. The pre-runs tower (T540) separates the light components (mainly water and traces of iso-butanal) from the crude alcohol. They are condensed and collected in the overhead drum (D541) which routes the organic phase back to the tower as reflux. From time to time a small portion has to be sent to the storage tank (TK-640) in order to control the light end components that may build up in the tower. The aqueous phase, which develops in the overhead drum (D541), is returned to the hydrogenation reactor (R520) (Van 2007).

The light-free bottom product stream passes to the finishing tower (T550), where purified iso-butanol is obtained as a liquid. The product is cooled and pumped to storage (TK650) in ISBL tankfarm. The overhead vapours are condensed and collected in the overhead drum (D551). Non-condensables are separated and sent to the flare. Pump (P551) forwards the liquid back to the finishing tower as reflux intermittently, but a small portion has to be sent to TK-640 to control the light components in the tower. The heavy ends (Oxo-oil 900) from the tower sump are also sent to TK-640 (Van 2007).

2.4 Process Simulation

Chemical process simulation aims to represent a process of chemical or physical transformation through a mathematic model that involves the calculation of mass and energy balances coupled with phase equilibrium and with transport and chemical kinetics equations. All this is made looking for the establishment (prediction) of the behavior of a process of known structure, in which some preliminary data of the equipment that constitute the process are known(Chaves et al. 2015).

Process simulators allow (Chaves et al. 2015):

- Predict the behavior of a process

- Analyze in a simultaneous way different cases, changing the values of main operating variables
- Optimize the operating conditions of new or existing plants
- Track a chemical plant during its whole useful life, in order to foresee extensions
- process improvements

2.4.1 Types of Process Simulators

Process simulators are classified according to the simulation strategy that they use to set the mathematical model that represents the process to simulate. The simulation strategy refers to the way in which the problem of the model solution is boarded. Generally, the strategy depends on the complexity of the model and the calculation mode. The first one, understood as the different existing possibilities, since linear to sophisticated models with equations of mass, energy and momentum transfer rates. The second, referred to the information (input variables) that is necessary to specify to solve the model in terms of the remaining information (output variables) (Chaves et al. 2015).

The subroutines of a process simulator are computer programs supplied initially with vectors containing the information corresponding to the feed streams of the process and some of its parameters. The subroutine takes the vectors, interprets the information, and looks for the appropriate model to solve the problem. The results are, basically, the product streams of the process. Thus, the subroutines permit working with two calculation modes in a process simulator (Chaves et al. 2015):

- Design mode: according to the required process conditions, a desired performance is used as starting point to find the process or equipment specifications that allow the accomplishment of those conditions (Chaves et al. 2015).
- Rating mode: according to some design specifications provided to the simulator, the performance of the process or equipment is evaluated to meet some specific conditions of the process (Chaves et al. 2015).

2.4.2 Aspen Plus and Aspen Hysys

Aspen Plus and Aspen Hysys are process simulators in steady state used in the prediction of the behaviour of a process or a set of unit operations, through the existing relationships between them. The relations and connections standing in the process determine, over the mass and energy balances, the phase and chemical equilibrium and the chemical transformation rates. In this way, it is possible to simulate the behaviour of existing or projected plants, with the objective of improving the design specifications or increase the profitability and efficiency of an operation in process.The main functions that can be found in this simulators are(Chaves et al. 2015):

- Generation of plots and tables
- Performing of sensitivity analysis and cases of study
- Sizing and rating of equipment
- Experimental data adjustment
- Analysis of pure components and mixtures properties
- Study of residue curve maps
- Process optimization
- Estimation and regression of physicochemical properties
- Dynamic analysis of processes

Aspen Plus is located between the group of simulators using the sequential strategy, in the same way as other simulators such as PRO II and CHEMCAD. Thus, it is composed by a group of simulation or program units (subroutines or models) represented through blocks and icons, to which the pertinent information must be provided to solve the mass and energy balances. However, it is important to mention that in the last versions of this simulator the possibility to work with the simultaneous or equation-oriented strategy has been included, allowing the modeling of systems and processes much more complex, highly integrated and with a high number of recycles(Chaves et al. 2015).

Aspen Hysys is a process simulator widely used in an industrial level, especially in conceptual design and detailed engineering, control, optimization and process monitoring stages in a project. The most important applications of Aspen Hysys correspond to the industries of oil and gas processing, refineries, and some industries of air separation. All these practices take advantage of this simulator architecture that permits the integration of the steady-state and dynamic models in an only unit. In this way, it is possible to bring together the stages of process design with the rigorous analysis of the dynamic behaviour and the control of the same, to evaluate in a direct way the effects that the decisions in the detailed design step have over the dynamic and controllability of the process(Chaves et al. 2015).

2.4.3 Physical Property Methods (Thermodynamic Model)

This essential first step will affect all subsequent tasks in developing accurate physical properties in your simulation. Indeed, the choice of the physical property models for a simulation can be one of the most important decisions for an engineer. Several factors need to be considered, and no single method can handle all systems. Table 2.5 lists some thermodynamic models available in simulators. The four factors that you should consider when choosing property methods are (Eric 1996):

- The nature of the properties of interest;
- The composition of the mixture;
- The pressure and temperature range; and
- The availability of parameters.

Table 2.5 Types of Thermodynamic Models

Equation-of-State Models	Activity Coefficient Models
Benedict-Webb Rubin(BWR) Lee Starling Hayden-O'Connell*	Electrolyte NRTL Flory-Huggins
Hydrogen-fluoride equation of state for	NRTL
hexamerization*	Scatchard-Hildebrand
Ideal gas law*	UNIQUAC
Lee-Kesler (LK)	UNIFAC
Lee-Kesler-Plocker	Van Laar
Peng-Robinson (PR) Perturbed-Hard-Chain	Wilson
Predictive SRK	Special Models
Redlich-Kwong (RK)	API sour-water method
Redlich-Kwong-Soave (RKS)	Braun K-10
RKS or PR with Wong-Sandler mixing rule	Chao-Seader
RKS or PR with modified-Huron-Vidal-2 mixing rule	Grayson-Streed
Sanchez-Lacombe for polymers	Kent-Eisenberg
* Not used for the liquid phase.	Steam Tables

Source: Eric (1996).

This is the area in which the most physical property work is focused in chemical engineering. Liquid/liquid equilibrium (LLE) also becomes important in processes such as solvent extraction and extractive distillation. Another critical consideration is pure-component and mixture enthalpy. Enthalpies and heat capacities are important for unit operations such as heat exchangers, condensers, distillation columns, and reactors(Eric 1996).

The universal quasi-chemical (UNIQUAC) equation, as introduced by Abrams and Prausnitz in 1975 expresses this dependency. It is widely used to calculate activities and activity coefficients, not only for ''liquid–liquid,'' but also for vapor-liquid equilibria. Since it only needs two adjustable parameters per binary pair and no ternary constants, it is a useful theory to describe the activity of multicomponent mixtures. The original UNIQUAC equation describes the fluid-phase equilibrium in molar terms, and it is widely used to simulate hydroformylation process with low pressure (Verhoef 2010). The chosen the appropriate physical method, the analysis showed that the process is suitable for the use of PR, SRK two physical methods to respond, using

WILSON, NRTL, UNIQUAC and other physical properties of the process of distillation simulation (Xue, Y. 2017).

2.5 Kinetics Equations of Aldehyde Hydrogenation

(OLDENBURG and RACE 1957) studied the kinetics of the vapor phase hydrogenation of three aldehydes (acetaldehyde, propionaldehyde, and *n*-butyraldehyde). Equations (2.6) and (2.7) represent the kinetics equations of aldehyde.

$$r_{\circ} = k\frac{P_{H}}{(P_{H})^{1/2}} \qquad 2.6$$

$$\ln k = -\frac{E}{RT} + \ln A \qquad 2.7$$

Where $\ln A_{acetaldehyde} = 4.94$

$\ln A_{propionaldehyde} = 5.17$

$\ln A_{n\text{-}butyraldehyde} = 5.31$

$E_{acetaldehyde} = 7890$ J/kmol

$E_{propionaldehyde} = 7810$ J/kmol

$E_{n\text{-}butyraldehyde} = 7739$ J/kmol

CHAPTER 3

MATERIALS AND METHODS

3.1 Introduction

The simulation of packed bed reactors for the production of butanol via hydrogenation of butanal has been carried out in the present study. The research activities are planned and arranged based on the procedure required as shown in Figure 3.1. The components involved in the process were first identified based on the main and side reactions. A suitable thermodynamic model has been chosen based on the interaction of these components. In addition, the kinetic equations for both the main and side reactions have obtained from the literature review. The thermodynamic model and kinetic equations incorporated together with the feed and reactor specifications into the packed bed reactor (PBR) model for the simulation of the hydrogenation of butanal to produce butanol. The simulation results generated from the computer simulator have been compared with an industrial plant data. In the case of the relative deviation exceeds 10% from the industrial plant data, the kinetic equation has been revised. Otherwise, the validated model has been used to execute. the investigation of the effect of different reactor schemes and operating conditions.

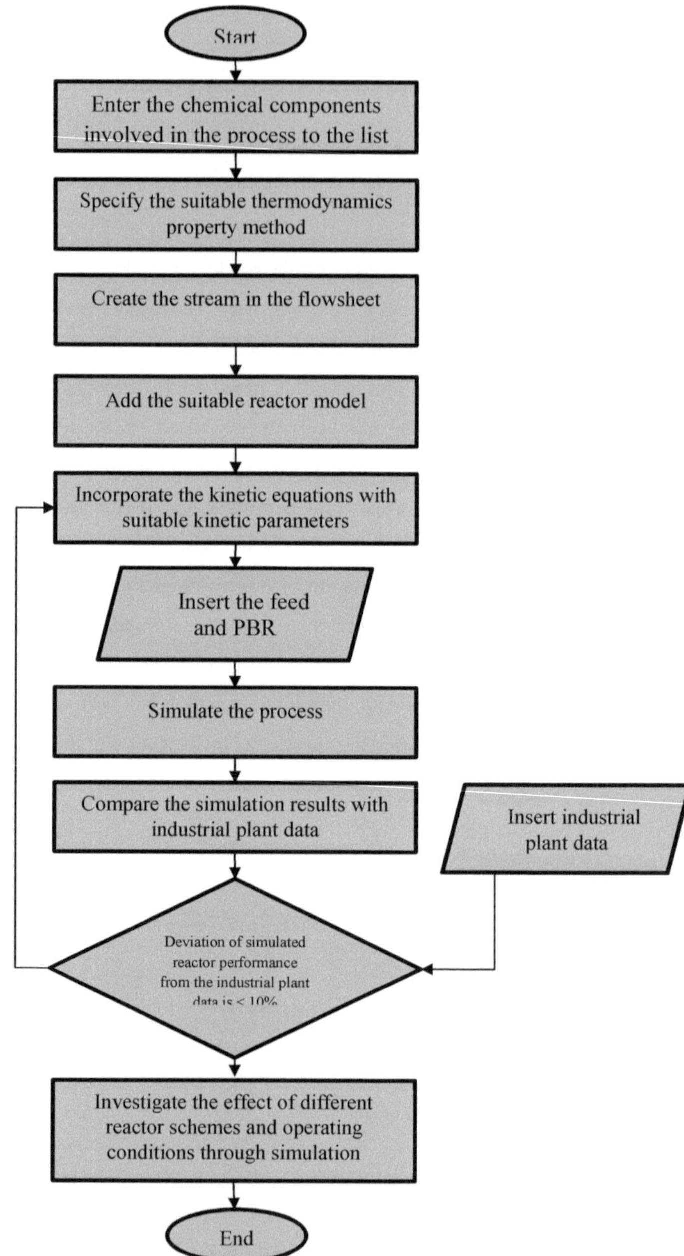

Figure 3.1 Process flow chart for the simulation study

3.2 Selection of suitable of thermodynamic model

Proper selection of thermodynamic models during process simulation is absolutely necessary as a starting point for accurate process simulation (Suppes 2002). Selection of appropriate equation of states or activity coefficient equations for phase equilibrium calculations has a lot of effects on precision and accuracy of the calculated results. It is obvious that any equation of state or activity coefficient models have some application limitations and one cannot use them for any systems at any temperature, pressure or compositions. Therefore, one has to select the appropriate thermodynamic models for modelling and simulations of a system before any calculations (Eliassi et al. 2008).

The Aspen Plus simulator software requires a correct property selection for the components in the system in order to run the simulation in a more accurate way. In the simulator software, there exists a "property method" selection tab which contains a collection of estimation methods to calculate several thermodynamic and transport parameters. Thermodynamic parameters are fugacity, enthalpy, entropy, Gibbs free energy and volume, whereas, for transport parameters, there are few such as viscosity, thermal conductivity, diffusion coefficient and surface tension. In the present study, the NRTL activity coefficient model has been chosen to evaluate the gas phase and liquid phase ideal system (Eliassi et al. 2008).

3.3 Modelling Approach

3.3.1 Packed bed reactor model

A fixed packed bed reactor consists of a compact, immobile stack of catalyst pellets within a generally vertical vessel. On macroscopic scales, the catalyst bed behaves as a porous media. The fixed beds are thus employed as continuous tubular reactors in which the reactive species in the mobile fluid (gas or liquid) phase are reacting over the catalyst surface (interior or exterior) in the stationary packed bed (Jakobsen 2008).

As with the PFR, the PBR is assumed to have no radial gradients in concentration, temperature, or reaction rate. The generalised mole balance on species A over catalyst weight AW results in the equation (Scott Fogler 2016).

$$In - Out + Generation = Accumulation \qquad (1)$$

$$F_{AW} - F_{A(W+\Delta W)} + r'_{A\Delta W} = 0$$

The differential form of the mole balance for a packed-bed reactor:

$$\frac{dF_A}{dW} = r'_A \tag{2}$$

If the pressure drop and catalyst decay are neglected,

$$W = \int_{F_A}^{F_{A0}} \frac{dF_A}{-r_A} \tag{3}$$

where W is the catalyst weight needed to reduce the entering molar flow rate of species A, F_{A0}, to a flow rate F_A.

The simulation of packed bed reactors for the production of butanol via hydrogenation of butanal was carried out using R-Plug in Aspen Plus. RPlug is a rigorous model for plug flow reactors which assumes that perfect mixing occurs in the radial direction and that no mixing occurs in the axial direction. RPlug can model one-, two-, or three-phase reactors. This model is useful when reaction kinetics is known, including reactions involving solids (Aspen Plus guideline 2016).

The process flow diagram of the original reactor configuration (one vapour phase reactor followed by two liquid phase reactors) is shown in Figure 3.2. The feed stream entered the first reactor with the total flow rate of 1506.26 kmol.hr^{-1}. It comprised of 0.703 mol% of H_2, 0.213 mol% of CH_4, 0.004 mol% of H_2O, 0.0.058 mol% of *n*-butanal, 0.017 mol% of i-butanal, 0.0.003 mol% of *n*-butanol, and 0.002 mol% of i-butanol. The reaction temperatures were 213°C, 159°C, and 167°C for the first, second and third reactor respectively. Whereas the reaction pressures for vapour phase and liquid phase reactors were 21 bar and 16 bar respectively.

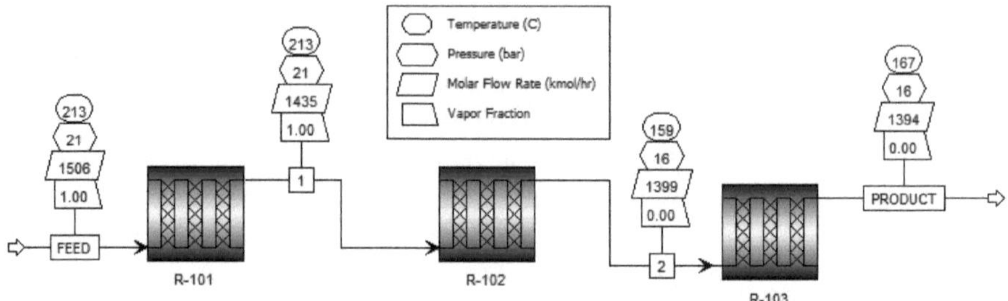

Figure 3.2 Original reactor configuration for butanal hydrogenation.

3.3.1.1 Kinetic model

Butanal hydrogenation

Eq.(4) and Eq.(5) show the chemical equation and kinetic model of *n*-butanal and *i*-butanal for the production of butanol through butanal hydrogenation, which is found as endothermic reaction.

$$C_4H_8O + H_2 \rightarrow C_4H_9OH \tag{4}$$

$$r_{bals} = \mathrm{A}\exp\left(-\frac{\mathrm{E}}{\mathrm{T}}\right)\mathrm{P_{bal}}^{0\cdot7}\,\mathrm{P_{H2}}^{0\cdot5} \tag{5}$$

Where r_{bals} = Butanal reaction rate (Moles/liter.hr)

A_g = 33.38 Moles/liter.bar

A_l = 1.59345×10^7 Moles/liter.bar

E_g = 2.9481×10^{10} KJ/kmol

E_l = 60520 KJ/kmol

P_{bal}= Partial pressure of butanal (bar)

P_{H2}= Partial pressure of hydrogen (bar)

Where "g" and "l" refer to gas and liquid phase reactions.

Side *Reaction*

Beside butanal hydrogenation, the elevated temperature in this process also promotes side products such as butyl butyrate. Butyl butyrate is produced from the hydrogenation of n-butanol, as shown in eq (6):

$$2C_4H_9OH \rightarrow C_8H_{16}O_2 + H_2 \tag{6}$$

The rate equation for the side reaction is shown in eq (7).

$$r_{byprod} = \mathrm{A}\exp\left(-\frac{\mathrm{E}}{\mathrm{T}}\right)\mathrm{P_{bal}}^{1\cdot8} \tag{7}$$

Where r_{byprod} = Byproduct reaction rate (Moles/liter.hr)

A_g = 3744.539 Moles/liter,bar

A_l = 1.92079×10^9 Moles/liter.bar

E_g = 4.9884×10^{10} KJ/kmol

E_l = 100973.53 KJ/kmol

P_{bal}= Partial pressure of butanal (bar)

P_{H2}= Partial pressure of hydrogen (bar)

3.4 Input relevant feed and PBR specifications values for the execution of computer simulation

Input values such as component species, feed flow rates, feed compositions, and specifications of PBRs such as pressure, temperature, molar flow rates, and other relevant data has entered to the Aspen Plus V9 simulator software to run the simulation, and the reactor diameter and length were fixed as 2.75 m and 10.92 m. The details of the input conditions for the reactors are shown in Table 3.1.

Table 3.1 Normal operating condtions of the hydrogenation reactors

T °C	P bar	N (kmol /hr)	V_0 (l /min)	H_2	CH_4	H_2O	IB	NB	IOH	NOH
120	16 - 21	1506.25	48319.7	0.703	0.213	0.004	0.017	0.058	0.002	0.003

Source: Zhang et al (2005).

3.5 Simulation results comparison and verification (with industrial plant data)

The simulation results generated were used to compare with the plant data. The main target is to prove the validity of the models used before it was adopted further to predict the effect of bypassing the liquid phase reactors in terms of product quality, yield, and conversion. The relative deviations of the simulated results from the plant data are calculated according to equation (3.10).

$$\text{Relative Deviation } (\%) = \frac{\dot{m}_{pp,plant} - \dot{m}_{pp,sim}}{\dot{m}_{pp,plant}} \times 100\% \qquad (3.10)$$

Where $\dot{m}_{pp,plant}$ = Plant mass flow rate of butanol

$\dot{m}_{pp,sim}$= Simulated mass flow rate of butanol

The relative deviation of the model (simulation result) from the industrial plant data is less than 10%, the model is accepted and proceeded to determine the best operating condition of the PBR in the original reactor configuration and after the modification made, since it is somehow comparable with the trend of the industrial plant data. However, if the relative deviation does exceed 10% from the industrial plant data, it is judged to be not comparable with the industrial plant trend. With relative deviation that exceeds 10%, the flow was proceeded back to incorporate the kinetic equation with another suitable kinetic parameters.

3.6 Conversion and yield calculations

Conversion of n-butanal and i-butanal and the yields of products have calculated by using equations (3.11) and (3.12) which are reported by (Scott Fogler, 2016).

$$X_A = \frac{F_{in} - F_{out}}{F_{in}} \times 100\% \quad 3.11$$

$$Y_i = \frac{F_i}{F_{i0} - F_i} \quad 3.12$$

Where A = Chemical species

Y_i = Overall yield

F_i = Mole flow rate of species *I* (mol/s)

3.7 Determination of best operating condition of the packed bed reactor

As long as the simulation results of a particular model succeeds the criteria by comparing with the plant data, that particular model was used to model the butanal hydrogenation reaction. Table 3.2 summarizes the operating parameters for the hydrogenation reactor of original reactor configuration and after modified it that have varied to obtain the best conversion.

Table 3.2 Operating parameters for the hydrogenation reactors

Parameter	Hydrogenation Unit
Reaction temperatures for vapor phase reactor.	135°C - 265°C
Reaction temperature for liquid phase reactors.	128°C - 196°C
Reaction pressure	16 – 21 bar

Source: Industrial plant data

CHAPTER 4

RESULTS AND DISCUSSION

4.1 Model validation

In order to validate the selected thermodynamic, reactor and kinetic models, the simulated results were compared with the industrial plant data obtained from a packed bed reactor. The present study has limited to the validation of the gas phase reactor due to the unavailability of plant data for the liquid phase reactor.

The gas phase reactor was operated at 125°C and 21 bar. The feed flow rate was 1506.25 kmol/hr with the composition of H_2, H_2O, CH_4, IB, NB, IOH, and NOH.

As shown in Table 4.1 The butanal hydrogenation process is well described by the kinetic parameters since most of the data points of all components except H_2O are within the marginal error lines of 20%. The H_2O outlet composition is not accurately predicted due to its very low absolute value compared to the other components (Chin et al. 2016). The *n*-butanal conversion was 99.77% comparing with in the literature which is 99.99%. While the conversion of i-butanal was 100% typically the same conversion obtained. The reaction temperature was fixed as 125°C and the reaction pressure was fixed as 21 bar, based on the literature operating conditions.

In view of the average deviation of less than 10%, the validity of the models was confirmed, and these models were used in the following simulation studies.

Table 4.1 Comparison of the simulation results with the industrial plant data

Industrial plant data		Simulation results	Relative Deviation (%)
Zhang et al. (2005)			
Outlet composition			
N ($Kmol.h^{-1}$)	1393.18	1393.48	0.0216
H_2	0.679	0.679	0
CH_4	0.230	0.230	0
H_2O	0.005	0.004	20
n-butanal	0	0.000143	0
i-butanal	0	0	0
n-butanol	0.065	0.065	0
i-butanol	0.021	0.021	0

4.2 Original Reactor Configuration

The simulation of a reactor configuration with three (3) packed bed reactors in series (one vapour phase reactor followed by two liquid phase reactors) was done. The effects of different operation conditions on the performance of the third reactor were investigated, and the corresponding simulation results are shown in the following sections.

4.2.1 Effect of reaction temperature

Based on the industrial plant data, the conversion of *n*-butanal in vapor phase reactor should achieve 60% and for the liquid phase reactors, the conversion of the *n*-butanal should be 85% for the first liquid phase reactor and 99.5% for the second liquid phase reactor.

Figure 4.1 presents the variations of reactants conversions and product yields versus reaction temperature of the last liquid phase reactor in the series. The reaction was carried out at pressure of 21 bar for the vapor phase reactor and 16 bar for the liquid phase reactors. The reaction temperatures of the vapour phase reactor and first liquid phase reactor were fixed at 213 °C and 158.5 °C respectively. The reactors diameter and length were 2.75 m and 10.92 m, respectively.

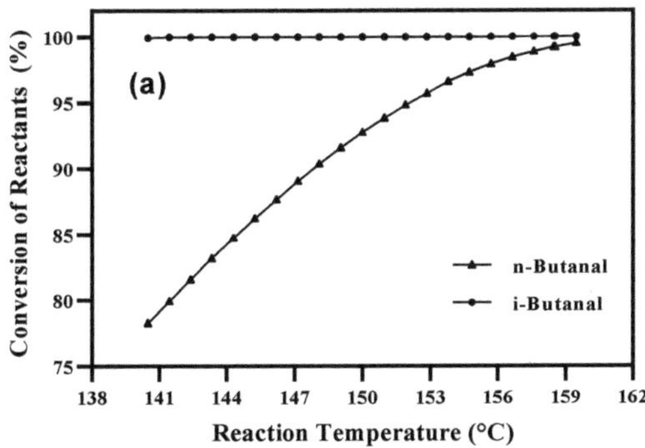

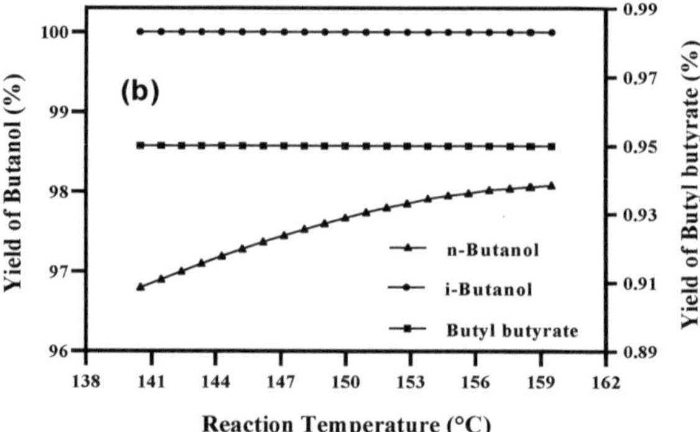

Figure 4.1 Effect of reaction temperature of the original configuration on (a) Conversion of reactants (b) Yield of products.

The conversion of *n*-butanal increased with the increase of temperature. Whilst the conversion of i-butanal was almost constant. With the increase in reaction temperature from 140.5 °C to 159.5 °C, *n*-butanal conversion increased from 78.24% to 99.48%. In addition, the yield of *n*-butanol increased from 96.8% to 98.08%, the yield of i-butanol was unvaried and remained at 100% and butyl butyrate yield was remained constant with a yield of 0.95%. The highest *n*-butanal conversion was obtained at temperature of 167 °C as 99.99%.

The theory of chemical reactions indicates that rates of reactions are generally enhanced by increase of temperature. However, experience shows that the maximum quantum of conversion of reactants to products does not increase monotonically. Indeed for a vast majority the maximum conversion reaches a maximum with respect to reaction temperature and subsequently diminishes (Smith et al., 2018).

The reason behind this phenomenon lies in the molecular processes that occur during a reaction. The rates of the forward and backward reactions both depend on temperature; however, an increase in temperature will, in general, have different impacts on the rates of each. Hence the extent of conversion at which they become identical will vary with temperature; this prompts a change in the equilibrium conversion. Reactions for which the conversion is 100% or nearly so are termed irreversible, while for those which never attains complete conversion are essentially reversible in nature (Smith et al., 2018).

4.2.2 Effect of reaction pressure

The effect of pressure on the conversion of reactants of the third reactor in the series can be seen in Figure 4.2. The operation temperatures of vapor phase reactor and two liquid phase reactors were fixed as 213 °C, 158.5 °C, and 159.5 °C. The reactors diameter and length were 2.75 m and 10.92 m, respectively.

The effect of pressure was investigated in the range of 16 to 21 bar for the second liquid phase reactor. It can be observed that when the pressure increased, the conversion of reactants was not affected negatively or a positively. The conversion of *n*-butanal and i-butanal only increased by 0.5% from 99.48% to 100% as the pressure was increased from 16 to 21 bar. In liquid-phase reactions, the concentration of reactants is insignificantly affected by even relatively large changes in the total pressure. Consequently, the effect of pressure can be totally ignored. Therefore, the pressure of 16 bar was chosen in the subsequent studies.

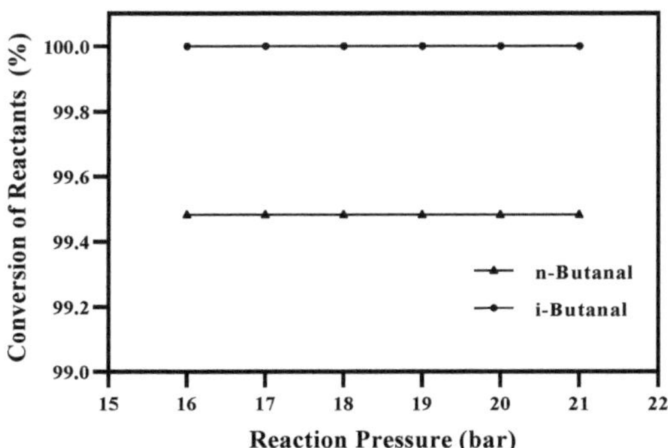

Figure 4.2 Effect of reaction pressure of the original configuration on the conversion of reactants.

4.2.3 Effect of reactor size

The effects of the reactor diameter and length were investigated of the third reactor in the series in the hydrogenation of butanal to produce butanol. The reaction was carried out at 213 °C, 158.5 °C, and 159.5 °C of reaction temperature and 21 bar, 16 bar, and 21 bar of reaction pressure for single vapor and two liquid phase reactors. The obtained results are shown in Figure 4.3.

The reactor volume changes with reactor length and reactor diameter. Reactor volume also affects the reaction performance. Therefore, sizing of the reactor is very important when high conversion and yield values desired. Therefore, the conversion of the reactants and the yield of products changes along the reactor. The reaction rate is a function of concentration; hence it also varies with reactor length. The reactor size affect the diffusion of the reactor, which is the spontaneous intermingling or mixing of atoms or molecules by random thermal motion. It gives rise to motion of the species relative to motion of the mixture. In the absence of other gradients (such as temperature, electric potential, or gravitational potential). molecules of a given species within a single phase will always diffuse from regions of higher concentrations to regions of lower concentrations (Scott Fogler 2016).

An increase in reactor length and diameter of the plug flow reactors caused a significant increase in the butanal conversion and in the yield of the products. The allowable L/D ratio is 4 to 10, therefore, the investigation of varies length and diameter was not exceeding the L/D ratio range through all the discussion. As seen in Figure 4.3, when the reactor diameter was increased from 1.25 to 2.75 m and reactor length was 10.92 m, *n*-butanal and i-butanal conversions increased from 24.32 to 99.48% and 35.38 to 100%. Whilst the yield of *n*-butanol, i-butanol, and butyl butyrate increased from 49.53 to 98.08%, 64.92 to 100%, and 0.33 to 0.95%. While the reactor length was increased from 10.92 m to 12 m, reactor diameter was 2.75 m, *n*-butanal reaction conversion values increased from 99.48 to 99.98%. While the conversion of *n*-butanal at 12.25 m of reactor length decreased to 99.96% and within increasing of reactor length to 12.5 m, lead to increasing of *n*-butanal conversion with maximum amount of 99.99%. While the conversion of i-butanal was remaining constant at 100%. While the yield of *n*-butanol decreases with reactor length, butyl butyrate yield increases. The highest conversion of *n*-butanal was obtained as 99.99% when the reactor diameter and reactor length was 2.75 m and 12.5 m, respectively.

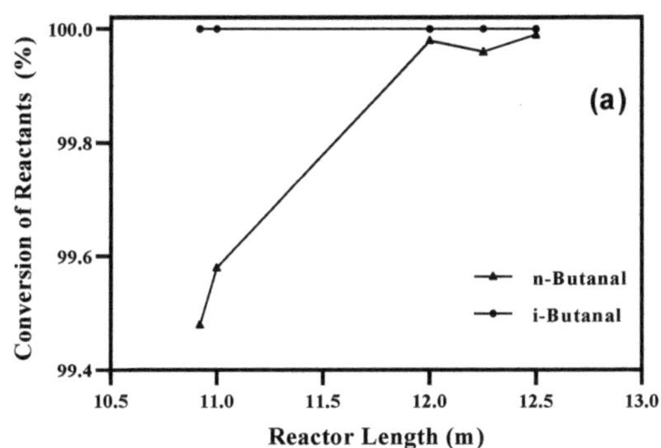
Conversion of Reactants (%)
(a)
n-Butanal
i-Butanal
Reactor Length (m)

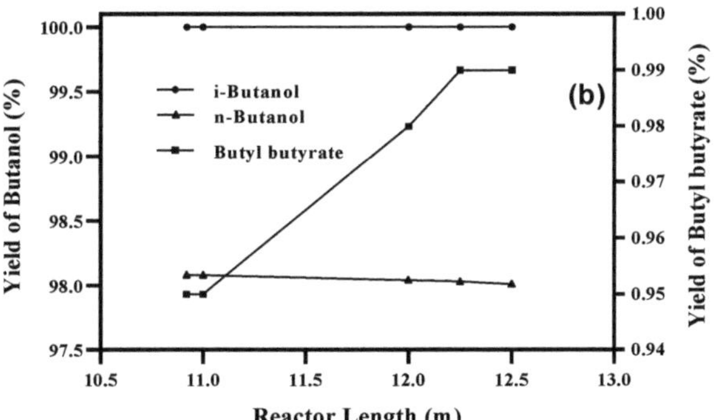
Yield of Butanol (%)
Yield of Butyl butyrate (%)
i-Butanol
n-Butanol
Butyl butyrate
(b)
Reactor Length (m)

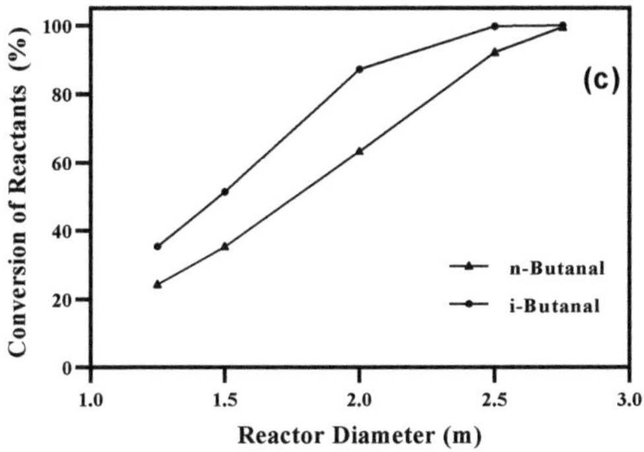
Conversion of Reactants (%)
(c)
n-Butanal
i-Butanal
Reactor Diameter (m)

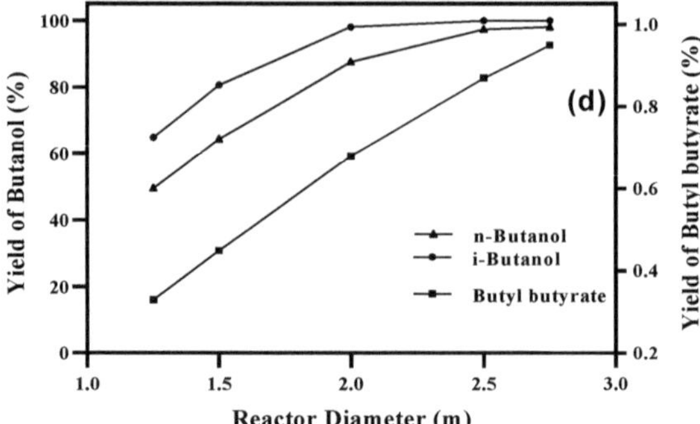

Figure 4.3 (a) Effect of reactor length of the original configuration on the conversion of reactants (b) Effect of reactor length on the yield of products (c) Effect of reactor diameter on the conversion of reactants (d) Effect of reactor diameter on the yield of products.

As shown in Table 4.2, the residence time of the second liquid phase reactor was calculated at varies reactor lengths and diameters. The reaction was carried out at 213 °C, 158.5 °C, and 159.5 °C in the first, second and third reactors respectively. Whereas the reaction pressures of these reactors are 21 bar, 16 bar, and 21 bar correspondingly. When the reactor diameter was increased from 1.25 to 2.75 m with the reactor length of 10.92 m, residence time increased from 5.83 to 28.8 minutes. While when the reactors length was increased from 10.92 to 12.5 m and reactor diameter fixed as 2.75m, residence time increased from 28.8 to 33 minutes. It is to conclude that the residence time of liquid phase reactors increases with increases of reactor length and diameter and resulting in increases of conversion values.

Table 4.2 The effect of reactor size of the original configuration on the reactor residence time

Reactor length fixed as 10.92 (m)			
Diameter (m)	Residence time (min)	n-butanal conversion (%)	i-butanal conversion (%)
1.25	5.8	24.32	35.38
1.5	8.5	35.36	51.29
2	15.2	63.22	87.15
2.5	23.8	92.17	99.7
2.75	28.8	99.48	100
Reactor diameter fixed as 2.75 (m)			
Length (m)	Residence time (min)	n-butanal conversion (%)	i-butanal conversion (%)
10.92	28.8	99.48	100
11	29	99.58	100
12	31.6	99.98	100
12.25	32.3	99.96	100
12.5	33	99.99	100

4.3 Modified Reactors Configuration

Simulation was also performed using the modified reactor configurations. Modified Configuration I comprised of two reactors in series, a vapour phase reactor followed by a liquid phase reactor, while Modified Configuration II constituted by solely the vapor phase reactor. The effects of variation of temperature, pressure and reactor size are reported as follows.

4.3.1 Modified Configuration I- Bypassing the Second Low Pressure Liquid Phase Hydrogenation Reactor

4.3.1.1 Effect of reaction temperature

As shown in Figure 4.4 the variations of reactants conversions and product yields versus reaction temperature of the second liquid phase in the series. The reaction was carried out at pressure of 21 bar for the vapor phase reactor and 16 bar for the liquid phase reactor. The reaction temperature of the vapor phase reactor was fixed as 255 °C and the reaction pressure were fixed as 21 bar and 16 bar for the vapor and liquid phase reactors. The reactors diameter and length were 2.75 m and 10.92 m, respectively.

The conversion of n-butanal increased with the increase of temperature from 77.32% to 99.45% and the conversion of i-butanal increased from 99.07 to 100 % with a range of temperature from 146.5 °C to 166.5 °. In addition, the yield of *n*-butanol increased from 93.39% to 96.7%, the yield of i-butanol is linear at 100% and butyl butyrate yield is remained constant with a yield of 1.61%. While the maximum *n*-butanal conversion was obtained at the temperature of 171 °C as 99.99%.

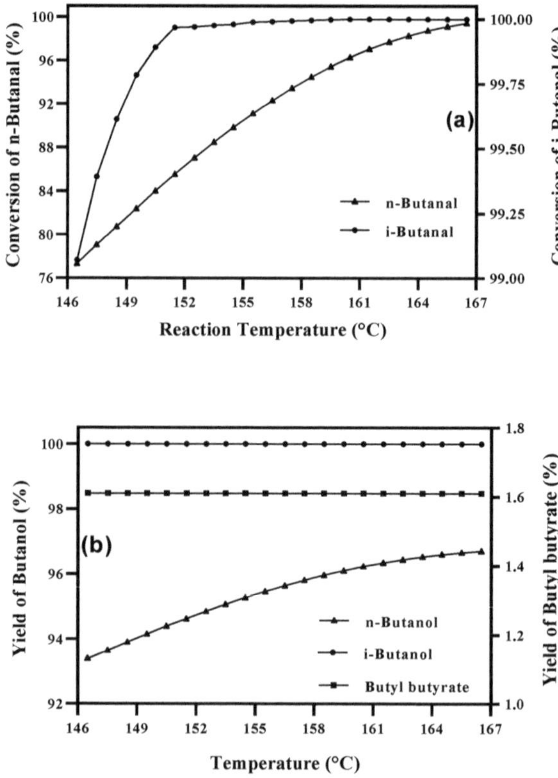

Figure 4.4 Effect of reaction temperature of the modified configuration I on (a) Conversion of reactants (b) Yield of products.

4.3.1.2 Effect of reaction pressure

The effect of pressure on the conversion of reactants can be seen in Figure 4.5. The operation conditions of the vapor phase reactor and liquid phaser reactor were fixed as 255°C and 166.5°C of the reaction temperature and the vapor phase reactor reaction pressure was fixed as 21 bar. The reactors diameter and length were fixed as 2.75 m and 10.92 m, respectively.

The effect of pressure was investigated in the range of 16 to 21 bar for the liquid phase reactor. It can be observed that when the pressure is increased, the conversion of reactants does not suffer a negative or a positive effect. Higher pressure or even lower pressure do not encourage butanal conversion. The maximum amount of reactants conversion can be achieved at any certain pressure. Therefore, the conversion of *n*-butanal and i-butanal are remained constant at 99.44% and 100% as the pressure was increased from 16 to 21 bar.

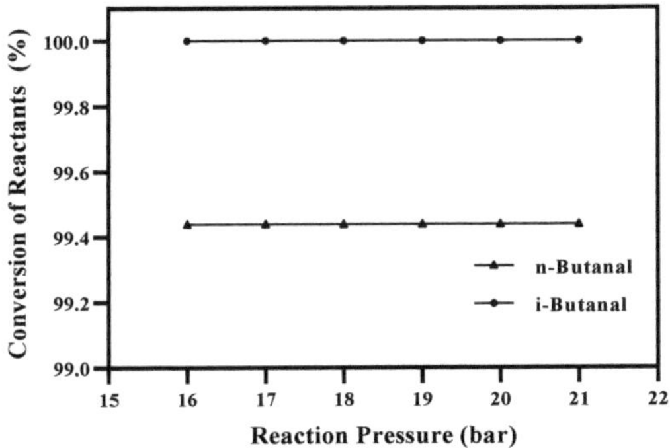

Figure 4.5 Effect of reaction pressure of the modified configuration I on the conversion of reactants.

4.3.1.3 Effect of reactor size

The reaction was carried out at 255 °C and 166.5 °C for the reaction temperature and 21 bar and 16 bar of reaction pressure for single vapor and liquid phase reactors. The obtained results are shown in Figure 4.6.

The reactor volume changes with reactor length and reactor diameter. Reactor volume also affects the reaction kinetics. Therefore, sizing of the reactor is very important when high conversion and yield values desired. Therefore, the conversion of the reactants and the yield of products changes along the reactor. The reaction rate is a function of concentration; hence it also varies with reactor length. An increase in reactor length and diameter of the plug flow reactors caused a significant increase in the butanal conversion and in the yield of the products. As seen in Figure 4.8, when the reactor diameter was increased from 1.25 to 2.75 m and reactor length was 10.92 m, *n*-butanal and i-butanal conversions increased from 30.07 to 99.44% and 42.41 to 100%. Whilst the yield of *n*-butanol, i-butanol, and butyl butyrate increased from 49.53 to 98.08%, 64.92 to 100%, and 0.78 to 1.61%. While the reactor length was increased from 10.92 m to 12 m, reactor diameter was 2.75 m, *n*-butanal reaction conversion values increased from 99.44 to 99.99%. While the conversion of *n*-butanal at 12.5 m of reactor length decreased to 99.98%. While the conversion of i-butanal was remaining constant at 100%. The yield of *n*-butanol increases at certain reactor lengths and decreases with others and butyl butyrate yield increases. The highest conversion of *n*-butanal was obtained as 99.98% when the reactor diameter and reactor length was 2.75 m and 12.25 m, respectively.

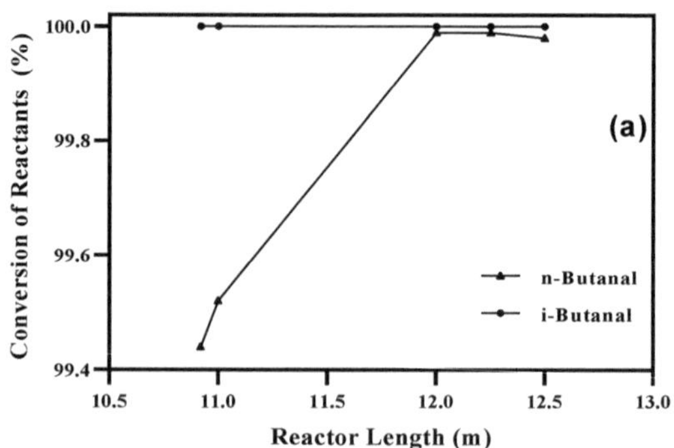

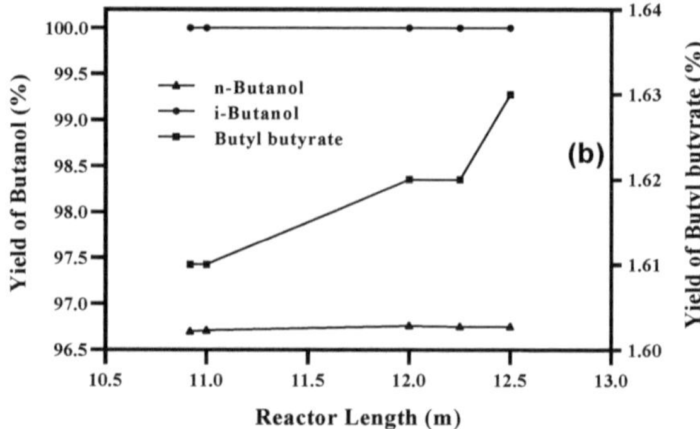

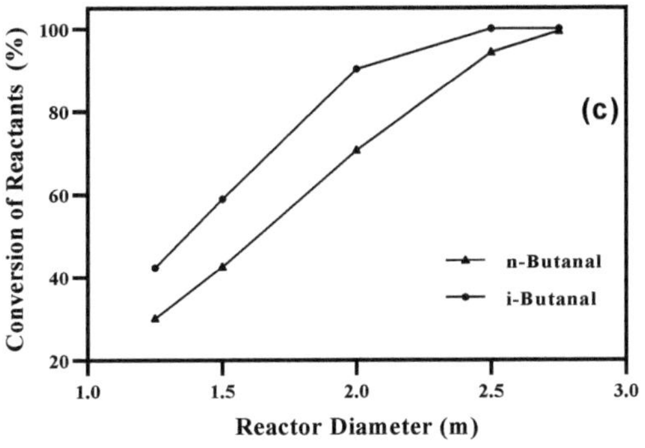

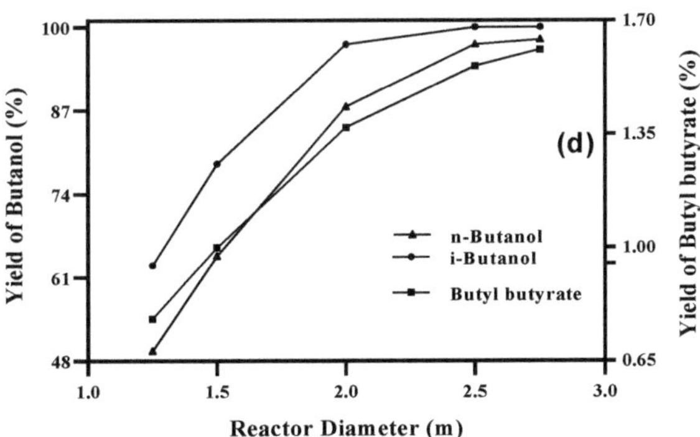

Figure 4.6 (a) Effect of reactor length of the modified configuration I on the conversion of reactants (b) Effect of reactor length on the yield of products (c) Effect of reactor diameter on the conversion of reactants (d) Effect of reactor diameter on the yield of products.

Table 4.3 shows the residence time of the liquid phase reactor at varies reactor lengths and diameters. The reaction was carried out at 255 °C, 166.5 °C of reaction temperature and 21 bar, 16 bar of reaction pressure for single vapor and liquid phase reactors. when the reactor diameter was increased from 1.25 to 2.75 m and reactor length was 10.92 m, residence time increases from 5.83 to 28.8 minutes. While when the reactors length was increased from 10.92 to 12.5 m and reactor diameter fixed as 2.75m, residence time increase from 28.8 to 33 minutes. We concluded that the residence time of liquid phase reactors increases with increases of reactor length and diameter and resulting in increasing of conversion values.

Table 4.3 The effect of reactor size of the modified configuration I on the reactor residence time

Reactor length fixed as 10.92 (m)			
Diameter (m)	Residence time (min)	*n*-butanal conversion (%)	i-butanal conversion (%)
1.25	5.8	30.07	42.41
1.5	8.4	42.55	58.95
2	15.2	70.66	90.27
2.5	23.8	94.31	99.98
2.75	28.8	99.44	100
Reactor diameter fixed as 2.75 (m)			
Length (m)	Residence time (min)	*n*-butanal conversion (%)	i-butanal conversion (%)
10.92	28.8	99.44	100
11	29	99.52	100
12	31.6	99.99	100
12.25	32.3	99.99	100
12.5	33	99.98	100

4.3.1.4 Benchmarking the simulation results of Modified Configuration I with that in the original reactor configuration

Based on the plant data, the targeted total conversion from the last reactor in the series should be 99.5%. In the Modified Configuration I, the total conversion of 99.44% from the first liquid phase reactor in the series obtained at reaction temperature of 166.5°C and reaction pressure of 16 bar. While the total conversion of the vapor phase reactor was fixed as 85% obtained at reaction temperature of 255°C and reaction pressure of 21 bar. Comparing with those in the original reactor configuration the total conversion of 99.48% from the second liquid phase reactor in the series obtained at reaction temperature of 159.5°C and reaction pressure of 16 bar. While the reactor length and diameter were fixed as 10.92 (m) and 2.75 (m) in the both cases. A more severe reaction temperature was required in the Modified Configuration I for achieving the targeted total conversion.

4.3.2 Modified Configuration II- Bypassing the first Low Pressure Liquid Phase Hydrogenation Reactor

4.3.2.1 Effect of reaction temperature

Figure 4.7 shows the variations of reactants conversions and product yields versus reaction temperature of a single vapor phase reactor. The reaction was carried out at pressure of 21 bar. The reactor diameter and length were 2.75 m and 10.92 m, respectively.

The total conversion of *n*-butanal and i-butanal increased with the increases of temperature. *n*-butanal conversion increased from 54.96% to 89.58% with industrial plant operation conditions limits. Whiles the conversion of i-butanal increased from 69.34 to 97.97 %. In addition, the yield of *n*-butanol increased from 53.31 to 86.06%, the yield of i-butanol increases from 69.34 to 97.97% and butyl butyrate yield is also increased from 0.82 to 1.76.% In the case of want to exceed the industrial plant operation conditions the maximum *n*-butanal conversion was obtained at the temperature of 285 °C and reaction pressure of 30 bar as 99.99%.

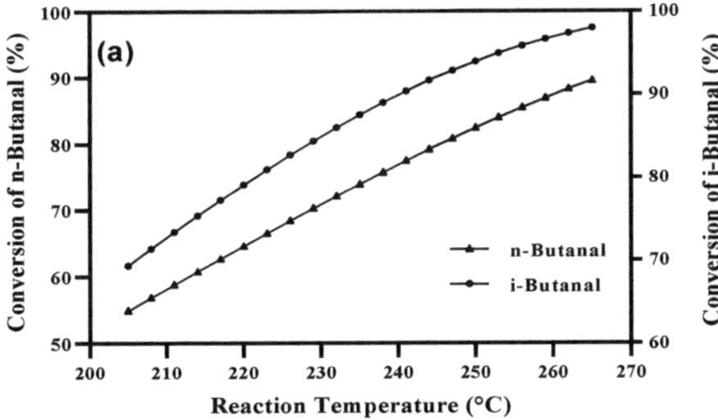

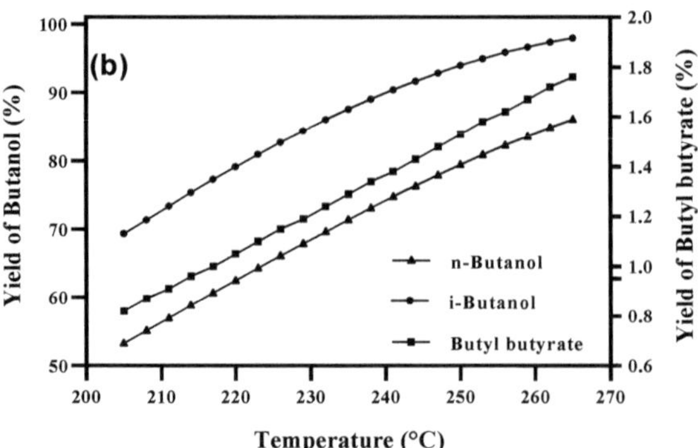

Figure 4.7 Effect of reaction temperature of the modified configuration II on (a) Conversion of reactants (b) Yield of products.

4.3.2.2 Effect of reaction pressure

The effect of pressure on the conversion of reactants and the yields of the products can be seen in Figure 4.8. The operation temperature of vapor phase reactor was fixed as 265°C. The reactors diameter and length were 2.75 m and 10.92 m, respectively.

The effect of pressure was investigated in the range of 16 to 21 bar for the vapor phase reaction. It can be observed that when the pressure is increased, the conversion of reactants or a positive effect. Higher pressure in vapor phase reaction encourages butanal conversion. . However. in gas-phase reactions, the concentration of the reacting species is proportional to the total pressure (Scott Fogler 2016). The maximum amount of reactants conversion can be achieved at 21 bar. Therefore, the conversion of *n*-butanal and i-butanal are increased from 76.79 to 89.58% and from 89.8 to 97.97% of i-butanal conversion. While the yields of the products increase from 73.9 to 86.06% of *n*-butanol yield and 89.77 to 97.97% of i-butanol, for the yield of butyl butyrate it increased from 1.45 to 1.76% as the pressure was increased from 16 to 21 bar.

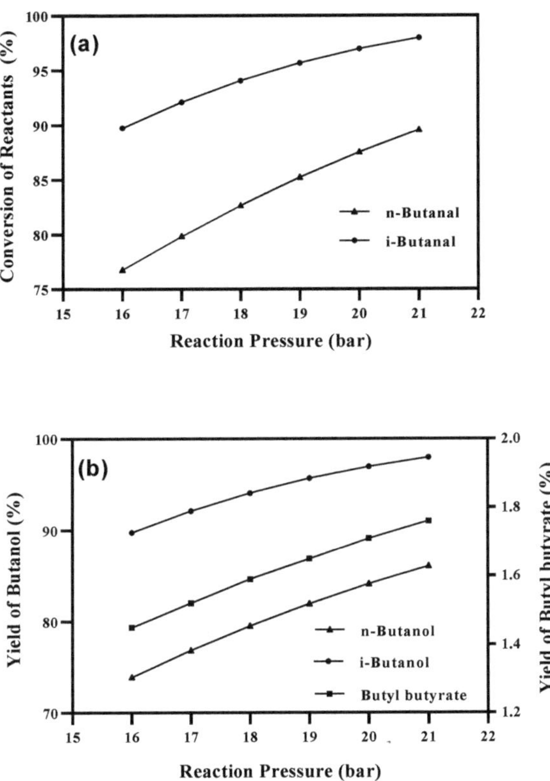

Figure 4.8 Effect of reaction pressure of the modified configuration II on (a) Conversion of reactants (b) Yield of products.

4.3.2.3 Effect of reactor size

The reaction was carried out at 265 °C and 21 bar of reaction pressure for vapor phase reaction. The obtained results are shown in Figure 4.9. The reactor volume changes with reactor length and reactor diameter. Reactor volume also affects the reaction kinetics. Therefore, sizing of the reactor is very important when high conversion and yield values desired. Therefore, the conversion of the reactants and the yield of products changes along the reactor. The reaction rate is a function of concentration; hence it also varies with reactor length. An increase in reactor length and diameter of the plug flow reactors caused a significant increase in the butanal conversion and in the yield of the products. As seen in Figure 4.11, when the reactor diameter was increased from 1.25 to 2.75 m and reactor length was 10.92 m, *n*-butanal and i-butanal conversions increased from 30.46 to 89.58% and 39.57 to 97.97%. Whilst the yield of *n*-butanol, i-butanol, and butyl butyrate increased from 28.62 to 86.06%, 39.57 to 97.97%, and 0.92 to 1.76%. While the reactor length was increased from 10.92 m to 12 m, reactor diameter was 2.75 m, *n*-butanal reaction conversion values increased from 89.58 to 93.66%. While the conversion of i-butanal was increasing from 97.97 to 99.45%. The yield of *n*-butanol and i-butanol increases from 86.06 to 90.12% and 97.97 to 99.45%. The butyl butyrate yield value is small with variation of reactor length, which is increased from 1.76 to 1.77%. The highest conversion of *n*-butanal was obtained as 93.65% when the reactor diameter and reactor length were 2.75 m and 12.5 m, respectively.

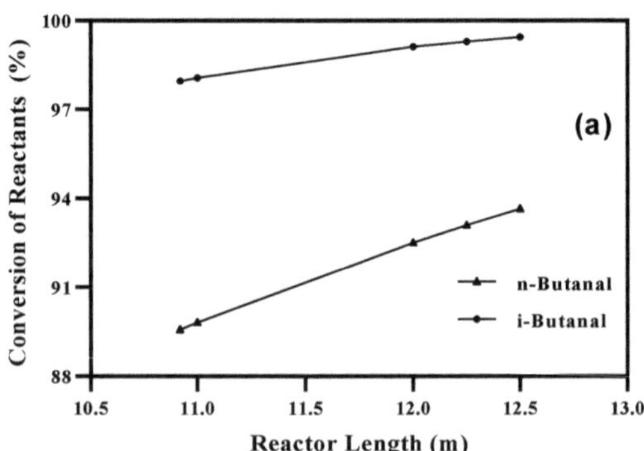

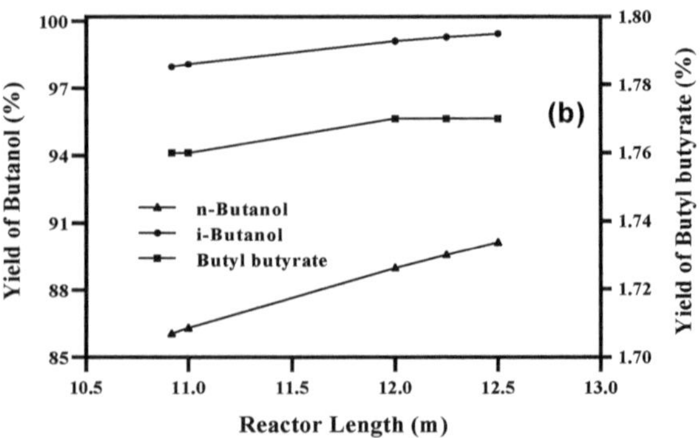

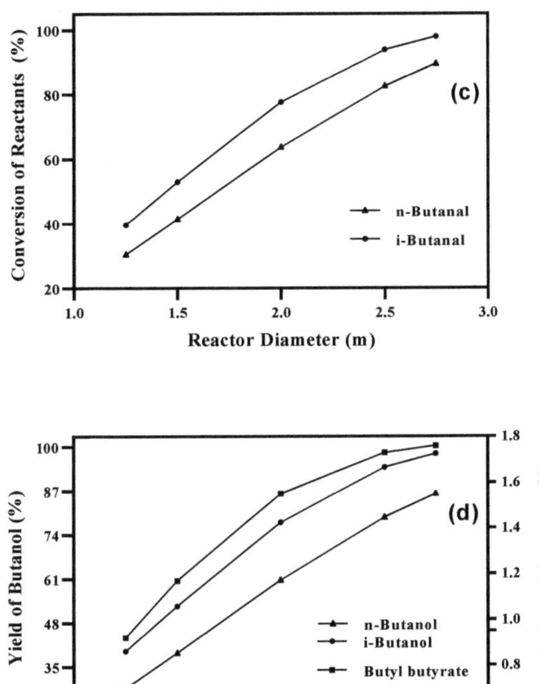

Figure 4.9 (a) Effect of reactor length of the modified configuration II on the conversion of reactants (b) Effect of reactor length on the yield of products (c) Effect of reactor diameter on the conversion of reactants (d) Effect of reactor diameter on the yield of products.

Table 4.4 shows the residence time of the vapor phase reaction at varies reactor lengths and diameters. The reaction was carried out at 265 °C of reaction temperature and 21 bar of reaction pressure. when the reactor diameter was increased from 1.25 to 2.75 m and reactor length was 10.92 m, residence time increases from 0.257 to 1.3 minutes. While when the reactors length was increased from 10.92 to 12.5 m and reactor diameter fixed as 2.75m, residence time increase from 1.3 to 1.5 minutes. We concluded that the residence time of vapor phase reaction increases with increases of reactor length and diameter and resulting in increases of conversion values.

Table 4.4 The effect of reactor size of the modified configuration II on the reactor residence time

Reactor length fixed as 10.92 (m)			
Diameter (m)	Residence time (min)	*n*-butanal conversion (%)	i-butanal conversion (%)
1.25	0.3	30.46	39.57
1.5	0.4	41.45	52.95
2	0.7	63.76	77.63
2.5	1.1	82.59	93.89
2.75	1.3	89.58	97.97
Reactor diameter fixed as 2.75 (m)			
Length (m)	Residence time (min)	*n*-butanal conversion (%)	i-butanal conversion (%)
10.92	1.3	89.58	97.97
11	1.3	89.82	98.08
12	1.4	92.52	99.12
12.25	1.5	93.12	99.29
12.5	1.5	93.66	99.45

4.3.2.4 Benchmarking the Comparison of the simulation results of Modified Configuration II with that in the original reactor configuration

In the Modified Configuration II, the total conversion from the vapor phase reactor can be obtained when the reaction temperature fixed as 285 °C and reaction pressure of 30 bar. Comparing with that in the original reactor configuration the total conversion value from the second liquid phase reactor in the series obtained at reaction temperature of 159.5°C and reaction pressure of 16 bar. While the reactor length and diameter were fixed as 10.92 (m) and 2.75 (m) in the both cases. The total *n*-butanal conversion in the Modified Configuration II was 99.99 % in the cast of the reactor operating conditions exceeded the industrial plant operation conditions . While in the original configuration was 99.48%. In relative to the original configuration and Modified Configuration I, the more severe operating temperature and pressure were needed to attain the desired total conversion.

4.4 Comparison of data with that in the literature

Table 4.5 compares the *n*-butanal conversion to *n*-butanol obtained from the present study with the findings of other researches about hydrogenation of butanal. In literature, hydrogenation of butanal was mostly carried out experimentally. Amomg the limited number of simulation studies found in literature, the effect of temperature was usually investigated.

Table 4.5 Comparison of conversion values with those in the literature

Temperature (°C)	Pressure (bar)	Catalyst type	Reactor type	Reaction Phase	Conversion of *n*-butanal (%)	References
139.85	3.1	SX-Rh SILP, and Shvo/SiO_2 catalysts	Packed bed reactor	Vapor	90	Hanna et al. (2014)
125	21	G-66 copper base catalyst	Tubular reactor	Vapor	99.99	Zhang et al. (2005)
210	5.8	Copper-zinc oxide catalysts	Fixed bed reactor	Vapor	99	Logsdon et al. (1989)
159.5	16	Copper-zinc oxide catalysts	Packed bed reactor	Liquid	99.48	This study
166.5	16	Copper-zinc oxide catalysts	Packed bed reactor	Liquid	99.45	This study
265	21	Copper-zinc oxide catalysts	Packed bed reactor	Vapor	89.58	This study

In this study, the conversion of *n*-butanal was mostly higher than in other studies in the different reaction conditions. Also, in the present studies, the high conversion of *n*-butanal was obtained under the higher temperature according to the literature. The differences between the conversion of values in our study and the values presented in the literature are related to reactor type, reactor sizing, and reaction phase. Packed bed reactor and NRTL thermodynamic model were chosen in this study. These are important points for the conversion values. The most appropriate reactor type, reactor sizing and thermodynamic model is resulted in high *n*-butanal conversion value. In addition, optimization of operation parameters is also important for the high conversion and yield values.

CHAPTER 5

CONCLUSIONS AND FUTURE RESEARCH DIRECTIONS

5.1 Introduction

The selected thermodynamic, reactor and kinetic models were validated by comparing the results with industrial plant data and the validated model has been used to execute. the investigation of the effect of different reactor configuration and operating conditions.

RPlug was employed for the reactor simulation. Sizing of the reactor was investigated in different diameter and length. The conversion of butanal increased with increasing reactor length. In the modified configuration I the total conversion of *n*-butanal was obtained at 166.5°C of the reaction temperature and 16 bar of the reaction pressure as 99.44% to 99.98% by variation of the reactor length from 10.92 m to 12.5 m. Reactor diameter was fixed at 2.75 m for the best of the reactor length. Therefore, reactor length is an important parameter for the reaction kinetics. The reactor diameter was also studied in the range of 1.25m to 2.75 m. The conversion was increased from 30.07% to 99.44% with diameter increment in addition, the yield of products increased. While the maximum *n*-butanal conversion of modified configuration II can be obtained when the reaction temperature fixed as 285 °C and reaction pressure of 30 bar with outside range of industrial plant operation conditions. Increased and by variation of the reactor length from 10.92 m to 12.5 m the total conversion of *n*-butanal increased from 89.58% to 93.65%. Reactor diameter was fixed at 2.75 m for the best of the reactor length. The reactor diameter was studied in the range of 1.25m to 2.75 m. The conversion was increased from 30.54% to 89.57% with diameter increment in addition, the yield of products increased. Comparing with that in the original reactor configuration the total conversion values obtained at reaction temperature of 159.5°C and reaction pressure of 16 bar. While the reactor length and diameter were fixed as 10.92 (m) and 2.75 (m) in the three cases. The reaction pressure also affected the reaction conversion and yield of products in the modified configuration II. The conversion of *n*-butanal increased from 76.79% to 89.57%. as the yield of *n*-butanol was raised from 73.9% to 86.06%. Therefore, the best operating conditions of the modified configuration I were 166.5°C of the reaction temperature and 16 bar of the reaction pressure with a total conversation of *n*-butanal as 99.44%. In addition, While the best operating conditions of the modified configuration II was obtained at 30 bar and 285 °C with a total conversion of *n*-butanal as 99.99% in the case of exceeded the industrial operating conditions limits.

Aspen Plus was successfully applied to identify the best operating conditions and reactor size of the last reactor in different reactor configurations. The best parameters obtained from the simulation study for butanal hydrogenation to produce butanol indicates considerable potential and courage for bypassing the liquid phase reactor.

5.2 Future research directions

Future investigations are necessary to develop the reaction kinetics for the main and side of the reactions, carry out controllability analysis to investigate the dynamic response of the process in the original and modified reactor configurations., and simulate the process with different thermodynamic models for comparison purposes, those are kinds of conclusions that can be drawn from this study.

REFERENCES

Agarwal, A.K., 2007. "Biofuels (alcohols and biodiesel) applications as fuels for internal combustion engines". Progress in Energy and Combustion Science, Vol. 33, No. 3, pp. 233 – 271.

Acres, Reviewed by Gary. 2007. 51 Platinum Metals Review *"Alcoholic Fuels."* http://openurl.ingenta.com/content/xref?genre=article&issn=0032-1400&volume=51&issue=1&spage=34.

Alviso, Dario Alberto. "Comparison between butanol and ethanol combusion kinetic molde through one-dimensional premixed".

Amiri, Hamid, and Keikhosro Karimi. 2018. "Pretreatment and Hydrolysis of Lignocellulosic Wastes for Butanol Production: Challenges and Perspectives." *Bioresource Technology* 270(July): 702–21. https://doi.org/10.1016/j.biortech.2018.08.117.

Barton W.E, Daugulis A.J., 1992, Evaluation of solvents for extractive butanol fermentation with Clostridium acetobutylicum and the use of poly(propylene glycol) 1200, Appl. Microbiol. Biotechnol. 36, 632-639.

Bergthorson, J.M. and Thomson, M.J., 2015. "A review of the combustion and emissions properties of advanced trans- portation biofuels and their impact on existing and future engines". Renewable and Sustainable Energy Reviews, Vol. 42, No. 0, pp. 1393 – 1417.

Brito, Marta, and Florinda Martins. 2017. "Life Cycle Assessment of Butanol Production." *Fuel* 208: 476–82. http://dx.doi.org/10.1016/j.fuel.2017.07.050.

Chen, C.-C., Mathias, P. M., & Orbey, H. (1999). Use of hydration and dissociation chemistries with the electrolyte–NRTL model. AIChE Journal, 45(7), 1576–1586. doi:10.1002/aic.690450719

Chaves, Iván Darío Gil et al. 2015. Process Analysis and Simulation in Chemical Engineering *Process Analysis and Simulation in Chemical Engineering.*

Cropley, JB, LM Burgess, and RA Loke. 1984. "The Optimal Design of a Reactor for the Hydrogenation of Butyraldehyde to Butanol." *ACS Symposium Series* 237: 255–70.

Chin, S. Y., Hisyam, A., & Prasetiawan, H. (2016). Modeling and Simulation Study of an Industrial Radial Moving Bed Reactor for Propane Dehydrogenation Process. International Journal of Chemical Reactor Engineering, 14(1). doi:10.1515/ijcre-2014-0148

Dahlbom S., Landgren H., Fransson P., 2011, Alternatives for Bio Butanol Production, Lunds Universitet feasibility study. Lund, Sweden.

Dürre P., 1998, New insights and novel developments in clostridial acetone/ butanol/ isopropanol fermentation, Appl. Microbiol. Biotechnol. 49, 639-648.

D. S. Sanders, D. M. Allen, and W. T. Sappenfield, Chem. Eng. Prog. 73(7), 40 (1977).

Eliassi, A, M M Rabiei, A Kargari, and H R Goodarzbod. "Selection of Appropriate Thermodynamic Equations for Simulation of Separation Columns of Vinyl Chloride Unit of Iran Bandar Imam Petrochemical Company (BIPC)." : 1–7.

Eric C. Carlson. 1996. "Dont Gamble With Physical Properties For Simulations." *Chemical Engineering Progress* (October): 35–46. https://pdfs.semanticscholar.org/a691/8690efb562b462410b4bb82a3ec58cd694bc.pdf.

Ezeji T.C., Qureshi N., Blaschek H.P., 2003, Production of acetone, butanol and ethanol by Clostridium beijerinckii BA101 and in situ recovery by gas stripping, World Journal of Microb. & Biotech. 19, 595–603.

Ezeji T.C., Qureshi N., Blaschek H.P., Karcher P.M., 2005, Improving performance of a gas stripping-based recovery system to remove butanol from Clostridium beijerinckii fermentation, Bioprocess Biosyst Eng 27, 207–214.

Ezeji, T C, N Qureshi, P Karcher, and H P Blaschek. 2006. "Production of Butanol from Corn." *Alcoholic Fuels* 1st(January 2015): 99–122.

Ezeji T.C., Qureshi N., Blaschek H.P., 2008, Fermentation of dried distillers' grains and soluble (DDGS) hydrolysates to solvents and value-added products by solventogenic clostridia, Bioresource Technology 99, 5232–5242.

Faanes, A. (2003). Controllability Analysis for Process and Control System Design (Doctoral dissertation). Retrieved from http://folk.ntnu.no/skoge/publications/thesis/2003_faanes/thesis_faanes_aug03.pdf.

Groot W. J., Van der Lans R.G.J.M., Luyben K.C.A.M., 1989, Batch and continuous butanol fermentations with free cells: integration with product recovery by gas-stripping, Applied Microb. & Biotech. 32, 305-308.

He, B.Q., Liu, M.B. and Zhao, H., 2015. "Comparison of combustion characteristics of n-butanol/ethanol - gasoline blends in a {HCCI} engine". Energy Conversion and Management, Vol. 95, No. 0, pp. 101 – 109.

Hanna, David G., Sankaranarayanapillai Shylesh, Pedro A. Parada, and Alexis T. Bell. 2014. "Hydrogenation of Butanal over Silica-Supported Shvo's Catalyst and Its Use for the Gas-Phase Conversion of Propene to Butanol via Tandem Hydroformylation and Hydrogenation." *Journal of Catalysis* 311: 52–58. http://dx.doi.org/10.1016/j.jcat.2013.11.012.

HANG Hai-tao,ZHANG Xin-rong,FANG Ding-ye~*(Department of Chemical Engineering, East China Universityof Science and Technology, Shanghai 200237, China); Mathematical Modeling of the Reactor Synthesizing Butanol from Butyraldehyde by Hydrogenation; Journal of East China University of Science and Technology; 2005-01.

J. L., R. L., J. M., & R. V. (1989). U.S. Patent No. US4876402A. Washington, DC: U.S. Patent and Trademark Office.

Jin, C., Yao, M., Liu, H., fon F. Lee, C. and Ji, J., 2011. "Progress in the production and application of n-butanol as a biofuel". Renewable and Sustainable Energy Reviews, Vol. 15, No. 8, pp. 4080 – 4106.

Ku, Jason T, Wiwik Simanjuntak, and Ethan I Lan. 2017. "Biotechnology for Biofuels Renewable Synthesis of n - Butyraldehyde from Glucose by Engineered Escherichia Coli." *Biotechnology for Biofuels*: 1–10. https://doi.org/10.1186/s13068-017-0978-7.

Lodi G., Pellegrini L.A., 2016, Recovery of butanol from abe fermentation broth by gas stripping, Chemical Engineering Transactions, 49, 13-18 DOI: 10.3303/CET1649003.

Liu Yuan, OXO Market Supply and Demand Forecast & Investment Economic Analysis , Vol.1, 2, 2012. ISSN: 2165 8226.

Maddox I.S., 1989, The Acetone-Butanol-Ethanol Fermentation: Recent Progress in Technology, Biotechnology and Genetic Engineering Reviews 7, 189-220.

Matthey, Johnson. 2017. "Industrial Low Pressure Hydroformylation : Forty-Five Years of Progress for the LP Oxo SM Process." (3): 246–56.

Matsumura M., Kataoka H., Sueki M., Araki K., 1988, Energy saving effect of pervaporation using oleyl alcohol liquid membrane in butanol purification, Bioprocess Engineering 2, 93-100.

M. Kopke, S. Noack, P. Durre (Eds.), The Past, Present, and Future of Biofuels - Biobutanol as Promising Alternative, InTech, Rijeka, Croatia, 2011.

Merriam, S, and W Voight. 1986. "United States Patent (19)." (19).

Michael W. Bradley, Norman Harris, Keith Turner. 1991. "(4) Patent Application Publication (212) Pub . No .: US 1991 / US 1991/5004845).

Niemistö, J., Saavalainen, P., Isomäki, R., Kolli, T., Huuhtanen, M. and Keiski, R.L., 2013. "Biobutanol production from biomass". In Biofuel Technologies, Springer, pp. 443 – 470.Qureshi, N et al. 2014. "Bioresource Technology Process Integration for Simultaneous Saccharification , Fermentation , and Recovery (SSFR): Production of Butanol from Corn Stover Using Clostridium Beijerinckii P260." *BIORESOURCE TECHNOLOGY* 154: 222–28. http://dx.doi.org/10.1016/j.biortech.2013.11.080.

Nielsen D.R., Prather K.J., 2009, In situ product recovery of n-butanol using polymeric resins, Biotechnology and Bioengineering 102, 811-821.

Oudshoorn A., Van der Wielen L.A.M., Straathof A.J.J., 2009, Assessment of Options for Selective 1-Butanol Recovery from Aqueous Solution, Ind. Eng. Chem. Res. 48, 7325-7336.

Qureshi N., Blaschek H.P., 2001, Recovery of butanol from fermentation broth by gas stripping, Renewable Energy 22, 557–564.

Qureshi N., Hughes S., Maddox I.S., Cotta M.A., 2005, Energy-efficient recovery of butanol from model solutions and fermentation broth by adsorption, Bioprocess and Biosystems Engineering 27, 215-222.

Oldenburg, C. C., & Rase, H. F. (1957). Kinetics of aldehyde hydrogenation: Vapor-phase flow system and supported nickel catalyst. AIChE Journal, 3(4), 462–466. doi:10.1002/aic.690030408

P. Durre, Biobutanol: an attractive biofuel, Biotechnol. J. 2 (12) (2007) 1525–1534.

Sadaka, Sammy. 2017. "Gasification , Producer Gas and Syngas."

Sauer, Michael. 2016. "Industrial Production of Acetone and Butanol by Fermentation — 100 Years Later." (May): 1–4.

Scott Fogler, H. 2016. "Elements of Chemical Reaction Engineering." *Chemical Engineering Science* 42(10): 2493. http://linkinghub.elsevier.com/retrieve/pii/0009250987801306.

Smith, J. M., et al. Introduction to Chemical Engineering Thermodynamics. McGraw-Hill Education, 2018.

Simulation, Introductionprocess. 2014. 35 *Introduction in Process Simulation 2 2.1.*

SRI international ,WP report, OXO, Version 2011.

Suppes, Galen J. 2002. "Selecting Thermodynamic Models for Process Simulation of Organic VLE and LLE Systems." http://people.clarkson.edu/~wwilcox/Design/thermod.pdf%5Cnpapers3://publication/uuid/AA20E4AE-5D15-41E1-B866-6DC1924F8C8B.

Tirado-Acevedo, Oscar, Mari S. Chinn, and Amy M. Grunden. 2010. 70 Advances in applied microbiology *Production of Biofuels from Synthesis Gas Using Microbial Catalysts.* 1st ed. Elsevier Inc. http://dx.doi.org/10.1016/S0065-2164(10)70002-2.

Tudor, By Richard, and Michael Ashley. 2007. "Enhancement of Industrial Hydroformylation Processes by the Adoption of Rhodium-Based Catalyst : Part I." (3): 116–26.

Tuţă, Emil Florin, and Grigore Bozga. 2011. "Performance Assessment by Simulation of a Gas-Recycle Oxosynthesis Plant with Propylene Recovery." *Industrial and Engineering Chemistry Research* 50(8): 4545–52.

Unlu, D., & Hilmioglu, N. D. (2019). Application of aspen plus to renewable hydrogen production from glycerol by steam reforming. International Journal of Hydrogen Energy. doi:10.1016/j.ijhydene.2019.02.106

Van, Eddy Theophyle Andrea. 2007. "(12) Patent Application Publication (10) Pub . No .: US 2007 / 0282134 A1." 1(19).

Vane L.M., 2008, Separation technologies for the recovery and dehydration of alcohols from fermentation broths, Biofuels Bioproducts and Biorefining 2, 553-588.

Verhoef, A., De Ridder, E., Degrève, J., & Van der Bruggen, B. (2010). Determination of activities in membrane processes: The UNIQUAC model expressed in mole and mass fractions. AIChE Journal, 57(7), 1889–1896. doi:10.1002/aic.12386

William Lee Jolly. 2018. "Hydrogen." Encyclopædia Britannica, inc. (February)

Xue, Y. (2017). Simulation and Optimization of Process for Production of Isobutanol by Syngas (Master thesis) Retrieved from https://epub.cnki.net/kns/brief/result.aspx?dbPrefix=CMFD

MIX
Papier aus verantwortungsvollen Quellen
Paper from responsible sources
FSC® C105338

Printed by Books on Demand GmbH, Norderstedt / Germany